T0255286

Lecture Notes in Computer Science

Vol. 49: Interactive Systems. Proceedings 1976. Edited by A. Blaser and C. Hackl. VI, 380 pages. 1976.

Vol. 50: A. C. Hartmann, A Concurrent Pascal Compiler for Minicomputers. VI, 119 pages. 1977.

Vol. 51: B. S. Garbow, Matrix Eigensystem Routines – Eispack Guide Extension. VIII, 343 pages. 1977.

Vol. 52: Automata, Languages and Programming. Fourth Colloquium, University of Turku, July 1977. Edited by A. Salomaa and M. Steinby. X, 569 pages. 1977.

Vol. 53: Mathematical Foundations of Computer Science. Proceedings 1977. Edited by J. Gruska. XII, 608 pages. 1977.

Vol. 54: Design and Implementation of Programming Languages. Proceedings 1976. Edited by J. H. Williams and D. A. Fisher. X, 496 pages. 1977.

Vol. 55: A. Gerbier, Mes premières constructions de programmes. XII, 256 pages. 1977.

Vol. 56: Fundamentals of Computation Theory. Proceedings 1977. Edited by M. Karpiński. XII, 542 pages. 1977.

Vol. 57: Portability of Numerical Software. Proceedings 1976. Edited by W. Cowell. VIII, 539 pages. 1977.

Vol. 58: M. J. O'Donnell, Computing in Systems Described by Equations. XIV, 111 pages. 1977.

Vol. 59: E. Hill, Jr., A Comparative Study of Very Large Data Bases. X, 140 pages. 1978.

Vol. 60: Operating Systems, An Advanced Course. Edited by R. Bayer, R. M. Graham, and G. Seegmüller. X, 593 pages. 1978.

Vol. 61: The Vienna Development Method: The Meta-Language. Edited by D. Bjørner and C. B. Jones. XVIII, 382 pages. 1978.

Vol. 62: Automata, Languages and Programming. Proceedings 1978. Edited by G. Ausiello and C. Böhm. VIII, 508 pages. 1978.

Vol. 63: Natural Language Communication with Computers. Edited by Leonard Bolc. VI, 292 pages. 1978.

Vol. 64: Mathematical Foundations of Computer Science. Proceedings 1978. Edited by J. Winkowski. X, 551 pages. 1978.

Vol. 65: Information Systems Methodology, Proceedings, 1978. Edited by G. Bracchi and P. C. Lockemann. XII, 696 pages. 1978.

Vol. 66: N. D. Jones and S. S. Muchnick, TEMPO: A Unified Treatment of Binding Time and Parameter Passing Concepts in Programming Languages. IX, 118 pages. 1978.

Vol. 67: Theoretical Computer Science, 4th GI Conference, Aachen, March 1979. Edited by K. Weihrauch. VII, 324 pages. 1979.

Vol. 68: D. Harel, First-Order Dynamic Logic. X, 133 pages. 1979.

Vol. 69: Program Construction. International Summer School. Edited by F. L. Bauer and M. Broy. VII, 651 pages. 1979.

Vol. 70: Semantics of Concurrent Computation. Proceedings 1979. Edited by G. Kahn. VI, 368 pages. 1979.

Vol. 71: Automata, Languages and Programming. Proceedings 1979. Edited by H. A. Maurer. IX, 684 pages. 1979.

Vol. 72: Symbolic and Algebraic Computation. Proceedings 1979. Edited by E. W. Ng. XV, 557 pages. 1979.

Vol. 73: Graph-Grammars and Their Application to Computer Science and Biology. Proceedings 1978. Edited by V. Claus, H. Ehrig and G. Rozenberg. VII, 477 pages. 1979.

Vol. 74: Mathematical Foundations of Computer Science. Proceedings 1979. Edited by J. Bečvář. IX, 580 pages. 1979.

Vol. 75: Mathematical Studies of Information Processing. Proceedings 1978. Edited by E. K. Blum, M. Paul and S. Takasu. VIII, 629 pages. 1979.

Vol. 76: Codes for Boundary-Value Problems in Ordinary Differential Equations. Proceedings 1978. Edited by B. Childs et al. VIII, 388 pages. 1979.

Vol. 77: G. V. Bochmann, Architecture of Distributed Computer Systems. VIII, 238 pages. 1979.

Vol. 78: M. Gordon, R. Milner and C. Wadsworth, Edinburgh LCF. VIII, 159 pages. 1979.

Vol. 79: Language Design and Programming Methodology. Proceedings, 1979. Edited by J. Tobias. IX, 255 pages. 1980.

Vol. 80: Pictorial Information Systems. Edited by S. K. Chang and K. S. Fu. IX, 445 pages. 1980.

Vol. 81: Data Base Techniques for Pictorial Applications. Proceedings, 1979. Edited by A. Blaser. XI, 599 pages. 1980.

Vol. 82: J. G. Sanderson, A Relational Theory of Computing. VI, 147 pages. 1980.

Vol. 83: International Symposium Programming. Proceedings, 1980. Edited by B. Robinet. VII, 341 pages. 1980.

Vol. 84: Net Theory and Applications. Proceedings, 1979. Edited by W. Brauer. XIII, 537 Seiten. 1980.

Vol. 85: Automata, Languages and Programming. Proceedings, 1980. Edited by J. de Bakker and J. van Leeuwen. VIII, 671 pages. 1980.

Vol. 86: Abstract Software Specifications. Proceedings, 1979. Edited by D. Bjørner. XIII, 567 pages. 1980

Vol. 87: 5th Conference on Automated Deduction. Proceedings, 1980. Edited by W. Bibel and R. Kowalski. VII, 385 pages. 1980.

Vol. 88: Mathematical Foundations of Computer Science 1980. Proceedings, 1980. Edited by P. Dembiński. VIII, 723 pages. 1980.

Vol. 89: Computer Aided Design - Modelling, Systems Engineering, CAD-Systems. Proceedings, 1980. Edited by J. Encarnacao. XIV, 461 pages. 1980.

Vol. 90: D. M. Sandford, Using Sophisticated Models in Resolution Theorem Proving.
XI, 239 pages. 1980

Vol. 91: D. Wood, Grammar and L Forms: An Introduction. IX, 314 pages. 1980.

Vol. 92: R. Milner, A Calculus of Communication Systems. VI, 171 pages. 1980.

Vol. 93: A. Nijholt, Context-Free Grammars: Covers, Normal Forms, and Parsing. VII, 253 pages. 1980.

Vol. 94: Semantics-Directed Compiler Generation. Proceedings, 1980. Edited by N. D. Jones. V, 489 pages. 1980.

Vol. 95: Ch. D. Marlin, Coroutines. XII, 246 pages. 1980.

Vol. 96: J. L. Peterson, Computer Programs for Spelling Correction: VI, 213 pages. 1980.

Vol. 97: S. Osaki and T. Nishio, Reliability Evaluation of Some Fault-Tolerant Computer Architectures. VI, 129 pages. 1980.

Vol. 98: Towards a Formal Description of Ada. Edited by D. Bjørner and O. N. Oest. XIV, 630 pages. 1980.

Vol. 99: I. Guessarian, Algebraic Semantics. XI, 158 pages. 1981.

Vol. 100: Graphtheoretic Concepts in Computer Science. Edited by H. Noltemeier. X, 403 pages. 1981.

Vol. 101: A. Thayse, Boolean Calculus of Differences. VII, 144 pages. 1981.

Vol. 102: J. H. Davenport, On the Integration of Algebraic Functions. 1–197 pages. 1981.

Vol. 103: H. Ledgard, A. Singer, J. Whiteside, Directions in Human Factors of Interactive Systems. VI, 190 pages. 1981.

Vol. 104: Theoretical Computer Science. Ed. by P. Deussen. VII, 261 pages. 1981.

Vol. 105: B. W. Lampson, M. Paul, H. J. Siegert, Distributed Systems – Architecture and Implementation. XIII, 510 pages. 1981.

Vol. 106: The Programming Language Ada. Reference Manual. X, 243 pages. 1981.

Lecture Notes in Computer Science

Edited by G. Goos and J. Hartmanis

142

Problems and Methodologies in Mathematical Software Production

International Seminar
Held at Sorrento, Italy, November 3–8, 1980

Edited by P.C. Messina and A. Murli

Springer-Verlag
Berlin Heidelberg New York 1982

Editors

Paul C. Messina
Mathematics and Computer Science Division
9700 South Cass Avenue, Argonne, IL 60439, USA

Almerico Murli
Istituto di Matematica "R.Caccioppoli", Università di Napoli
Via Mezzocannone 16, 80134 Napoli, Italy

CR Subject Classifications (1980): 2.41, 2.43, 4.43, 4.6, 5.0, 5.11, 5.16

ISBN 3-540-11603-6 Springer-Verlag Berlin Heidelberg New York
ISBN 0-387-11603-6 Springer-Verlag New York Heidelberg Berlin

Printed in Germany

Printing and binding: Beltz Offsetdruck, Hemsbach/Bergstr.
2145/3140-543210

Preface

In November, 1980 a week-long International Seminar on Problems and Methodologies in Mathematical Software Production was held in Sorrento, Italy.

The Seminar was organized and supported by the National Committee for the Mathematical Sciences and the National Group for Mathematical Informatics, both of which are part of Italy's National Council for Research.

Additional Sponsors for the seminar were Università di Napoli, AICA, Informatica Campania (Italsiel).

The Seminar was intended for university and industrial researchers with the following aims: to provide an opportunity for discussions among researchers and some of the well-known experts in the field of Mathematical Software; and to stimulate cooperation between industry and mathematical research.

The principal topics in the seminar were:

- fundamental concepts of computational software and the influence of computer architecture on the design of software;
- operational aspects of establishing, developing and maintaining mathematical software libraries;
- portability, transportability and software tools for numerical software;
- methods for evaluating numerical software;
- machine arithmetic and program correctness criteria.

Their presentations are collected in this book.

On the last day of the Seminar the speakers took part in a panel discussion. The discussion concentrated on current research directions in Mathematical Software; prospects for the future; and evolutionary trends now discernible, including the probable effects of new computer hardware architecture and technology.

A written version of the panel discussion appears at page 254.

P.C. Messina and A. Murli

Naples, 1981

SPEAKERS

William J. Cody, Jr.: Applied Mathematics Division, Argonne National Laboratory, 9700 South Cass Avenue, Argonne, IL 60439, USA.

Theodorus J. Dekker: Universiteit van Amsterdam, Subfaculteit Wiskunde/Department of Mathematics, Roetersstraat 15, 1018 WB Amsterdam.

Brian Ford: Numerical Algorithms Group Limited, NAG Central Office, Mayfield House, 256 Banbury Road, Oxford OX2 7DE.

Morven Gentleman: Computer Science Department, University of Waterloo, Waterloo N2L 3G1, Canada.

James N. Lyness: Applied Mathematics Division, Argonne National Laboratory, 9700 South Cass Avenue, Argonne, IL 60439, USA.

Paul C. Messina: Applied Mathematics Division, Argonne National Laboratory, 9700 South Cass Avenue, Argonne, IL 60439, USA.

CONTENTS

Basic Concepts for Computational Software

W. J. Cody*

Applied Mathematics Division
Argonne National Laboratory
Argonne, Illinois 60439

1. Introduction. The goal of research in computational mathematics is to supply the numerical tools needed by researchers in all fields. Development of these tools traditionally proceeds in an orderly fashion from the discovery of new numerical methods to their incorporation into computational algorithms addressing a particular problem set. But the job is not finished until these algorithms have been implemented in computer programs, or software. This last step, the implementation of computational algorithms in software, introduces new considerations that are largely independent of previous work on the algorithms.

This paper is a survey of those concepts that are fundamental to the task of implementing numerical algorithms. In the next section we define computational software and describe its relation to algorithms and activities in numerical analysis. In section 3 we introduce important details of the hardware environment in which computational software is written and used. In section 4 we define and discuss the concepts of reliability, robustness and transportability. In the last section we show how all these things affect the design of an algorithm and its implementation into software. Throughout the paper, simple examples involving elementary functions and hand calculators illustrate the ideas discussed. An additional example is presented in an Appendix. Finally there is a bibliography of cited and related, but uncited references.

* This work was supported by the Applied Mathematical Sciences Research Program (KC-04-02) of the Office of Energy Research of the U.S. Department of Energy under Contract W-31-109-Eng-38.

Much of the material in this paper has appeared before, but this is the first time it has been gathered together into one coherent body. In particular, see our earlier papers cited in the bibliography. Many individuals have contributed to our ideas without necessarily agreeing with them. Although available space precludes our mentioning everyone, we must acknowledge the influence of our colleagues at Argonne National Laboratory, and of W. Kahan and the late H. Kuki.

2. Computational Software. There are several distinct levels of activity in the development of computational software. The work done at a particular level is still somewhat independent of work done at other levels, although the differences between levels are becoming less distinct as we gain more experience in preparing software.

For purposes of discussion we will identify work at three levels. The first is the development of theoretical mathematical or numerical methods for solving a particular type of problem. The emphasis at this level is on theoretical considerations such as error analysis and proofs of convergence, although there is a growing awareness among theoreticians of the practical implications of the computer environment for their work. Crude computer programs are often written to verify that the methods under investigation work in some sense, but programs are not the primary result of this effort. Those that are written are certainly not computational software as we will define it.

The second level of activity is the combination of one or more theoretical methods into a practical computational algorithm specifically designed to solve a particular problem on a computer. This is not as easy a process as it sounds. The quadratic formula, for example, is a theoretical method for finding the roots of a quadratic polynomial equation. But Forsythe [15] has shown that in a computer environment various reformulations of the basic formula are necessary, depending on the particular coefficients, to determine the roots accurately.

The end product of this second level of activity is often considered to be a computer program, especially when the work is published in an algebraic programming language in a professional journal. However, we will use the formal term <u>algorithm</u> to describe results at this level, and reserve the formal term <u>program</u> to describe an item of <u>computational</u> <u>software</u>. We define computational software as running, documented implementations of algorithms in specific computer environments. This distinction is sometimes subtle and hard to defend, but it is important for the discussions that follow.

The key word in our definition of computational software is 'implementation'. Algorithms in the literature do not solve problems on computers; only implementations of algorithms solve problems. A good computer program must be based on a good algorithm, of course, but a good algorithm does not guarantee a good computer program. An improper implementation of even a superb algorithm may perform so poorly as to be useless, unfairly besmirching the reputation of the algorithm. Thus we must pay attention to the details of implementation. This introduces the third, and perhaps most important, level of activity: the implementation of algorithms to produce computational software. Here, especially, proper consideration of the environment is essential to success.

We motivate our concerns for the environment and for the design and implementation of algorithms by experimenting with some simple programs. Consider the evaluation of the magnitude of a complex number $z = x + iy$, where $i = \sqrt{(-1)}$. By definition,

$$|z| = \sqrt{(x^2 + y^2)}.$$

This definition corresponds to the theoretical method for determining $|z|$, and suggests the obvious algorithm of squaring the components, summing them, and extracting the square root. Suppose we implement that algorithm on a programmable hand calculator, a TI 59, say, and run it with $x = 3$ and $y = 4$. Then we find that $|z| = 5$, which suggests that we have implemented the algorithm correctly. Now suppose we scale the original data by 10^{-50}, and try again. This time the calculator stops with a blinking display of '1.-99'. Clearly there is an error of some sort in the calculation. Even though the data and correct result are all representable within the machine, the program is unable to return the correct result. At least the blinking display warns of trouble.

When we try the same experiment on an HP 34C calculator, we obtain the result '4.00-50' for the scaled problem. The result is again in error, but it is different from the previous one. Even worse, there is no indication of trouble this time. We conclude that the obvious algorithm is not necessarily the best algorithm, and that implementations on different machines behave differently. This conclusion reinforces our decision to distinguish between the algorithm and its implementation.

Each of these calculators contains a built-in program, accessible with a simple keystroke, for conversion between rectangular and polar coordinates. Such a program automatically determines $|z|$ as the radial coordinate. If we use the conversion program on a TI 59 with the scaled test data, we obtain the puzzling result '4.472136-50', again with no indication of error. This is more serious than before because we have used a 'system program' that is supposed to be correct, not one that we wrote. When systems programs contain errors, especially programs built into calculators, they are hard to correct. Happily, the conversion program built into the HP 34C returns the correct result '5.00-50'.

The results of this demonstration are not reassuring. We have tested four different programs for calculating $|z|$. Each program gives a different result for our test problem, only one of which is correct. The programs must all be based somehow on the definition of $|z|$ that we gave above, but one of them apparently modifies the obvious algorithm in some way to produce correct results when the others cannot. Therefore, the definition must lead to different computational algorithms, incidentally motivating the second level of activity described above. Only one of the incorrect results is accompanied by an error indication, and that seems to be accidental. We ought to agree that programs returning incorrect results without warning are dangerous. Therefore, there is more to a good program than the implementation of mathematical equations. Evidently good computational software has certain non-mathematical attributes that enhance its utility. We have also seen that implementations of the same algorithm on different machines can lead to dramatically different results. It is reasonable to assume that those results depend in some way on the host environment. Therefore, not only is there a difference between the algorithm and its implementation, but if we want uniform performance across machines, the implementations must take the environment into consideration. We will return to this example later after we have discussed the environment and software attributes.

3. The Environment. The computational environment is that milieu in which software exists. It includes the computer hardware on which programs are to execute as well as supportive software such as operating systems, compilers and libraries. We will delay a discussion of the supportive software until the next section and concentrate here on the hardware environment.

The floating-point arithmetic system includes both the scheme for representing data in the machine and the arithmetic engine that operates on that data. Ideally the arithmetic system should be 'clean' and free of anomalous behavior. Unfortunately, that is seldom the case. The essential differences between a floating-point number system and the real number system are related to the finite range and significance, hence the finite population of the floating-point system. Those properties of the real numbers not contradicting the finite character of the floating-point system ought to persist in that system. When they do not, the system is deficient and computational software needs to be carefully written.

For example, it is reasonable to expect that

$$1.0 * X = X,$$

$$X * Y = Y * X,$$

and

$$X + X = 2.0 * X$$

for all floating-point X and Y, because all these properties are independent of finite precision and range. It is unreasonable to expect that

$$X * (1.0/X) = 1.0$$

exactly for all non-zero X, because the intermediate result 1.0/X may require infinite precision for exact representation. There exist machines on which none of these properties hold. Worse yet, on some machines there exist floating-point numbers X such that X > 0.0, but

$$1.0 * X = 0.0$$

or

$$X + X = 0.0.$$

Such behavior boggles the imagination; it is contrary to all mathematical law. Fortunately, behavior this bad usually exists only on the fringes of an arithmetic system and involves numbers that are not ordinarily encountered in normal computation. Nevertheless, good computational software must be written defensively to guarantee that the computation will not be destroyed should it encounter such bizarre behavior.

Almost all the peculiarities of arithmetic systems are traceable to a few design parameters. To simplify our discussion we will assume that floating-point numbers are represented in a normalized sign-magnitude representation. That is, we assume that

$$X = \pm B^e f,$$

where $1/B \le f < 1$ unless $X = 0.0$, B is the <u>base</u> or <u>radix</u> of the representation, e is a signed exponent and f is the <u>significand</u> containing, say, b significant base-B digits. <u>Normalization</u> refers to the lower bound on f. Hand calculators can be thought of as decimal machines, and there is at least one base-3 machine, but usually the radix is a small integer power of 2, e.g., B = 2, 4, 8 or 16, because most computers are really binary.

The radix of the representation can cause problems. Consider the case of hexadecimal, or base-16 representation. In such a system the significand of a floating-point number contains b hexadecimal digits of 4 bits each, the first digit being non-zero if X is non-zero. Thus f contains $t = 4b$ bits with at least one of the first four bits being non-zero. The number of significant bits in f varies between t and t-3 as the magnitude of f varies (see Table I). There is almost one decimal place less significance in numbers with three leading zero bits than in numbers with no leading zero bits. This phenomenon is known as <u>wobbling precision</u>. Because floating-point numbers are logarithmically distributed [21], each of the four possible normalizations is equally likely to occur. On the average, 1/4 of all hexadecimal floating-point numbers have the worst normalization and contain only t-3 significant bits. We might expect that normal computational error would be larger in such a system than it would be in a corresponding binary system, but this is not necessarily the case (see [22]).

However, wobbling precision can impose an accuracy penalty on the unwary. One popular scheme for computing the tangent of an arbitrary argument X, for example, involves computing with the reduced argument w where

$$w = (4/\pi)X - N$$

Table I

Significance of Hexadecimal Fractions

f	Binary representation of f	Significant bits in f
$1/2 \le f < 1$	.1xxx...	t
$1/4 \le f < 1/2$	.01xx...	t-1
$1/8 \le f < 1/4$	.001x...	t-2
$1/16 \le f < 1/8$	.0001...	t-3

and N is an appropriate integer. We will see later that this computation is inaccurate for other reasons, but consider for now the contribution of the constant

$$4/\pi = 1.27\ldots = 16^1(1/16 + \ldots).$$

The hexadecimal representation of this number has three leading zero bits, hence at most t-3 of the bits in its product with X can be significant regardless of the normalizations of X and of the product. On the other hand, the related constant

$$\pi/4 = 0.785\ldots = 16^0(12/16 + \ldots)$$

contains no leading zero bits. Replacing the argument reduction involving $4/\pi$ with the scheme

$$w = X/(\pi/4) - N$$

increases the potential accuracy of w by almost one decimal place. Note that the first scheme is often preferable to the second on binary machines where there is no wobbling precision and multiplication is usually faster than division. This is a case where the details of the implementation should change from one environment to another to achieve maximum performance.

The major design parameters for the arithmetic engine that concern us are the mode of rounding, the presence of guard digits, the order of intermediate steps in an arithmetic operation, and the detection of out-of-bounds results. In a typical arithmetic operation the first step is to separate the exponent and significand of each operand for separate handling in later steps. We can think of the significands as being loaded into active arithmetic registers and the exponents as being carried as integers in other registers. Depending on the operation and the operands, some preliminary shifting of an operand and corresponding adjustment of an exponent may be necessary at this point. Next the desired operation is carried out using fixed-point arithmetic on the significands and integer arithmetic on the exponents. The intermediate significand is then rounded and normalized, although not necessarily in that order. Normalization, when it occurs, also requires an adjustment of the intermediate exponent. Finally, the result is reassembled from its components. Somewhere in this process any out-of-bounds results are detected.

We illustrate with multiplication. Let X and Y be two floating point numbers,

$$X = B^{e}f$$

and

$$Y = B^{e'}f',$$

where f = .xxx... , f' = .yyy... , and there are b base-B digits in f and f'. After separating the exponents and significands, an intermediate product is formed with exponent a = e+e', and significand g = f * f'. Note that a total of 2b base-B digits is necessary to represent the product f * f' exactly, although only b digits will be retained in the final result. If more than b digits of the product are developed in the arithmetic register and participate in subsequent renormalization and rounding, then we say there are <u>guard digits</u>. Otherwise only the first b digits of g are developed and there are no guard digits.

Because f and f' each lie in the interval [1/B, 1), the intermediate product g must lie in $[1/B^2, 1)$. Thus the first digit of g may be a zero, i.e., g may not be normalized. In that case the intermediate value of g must be shifted one digit to the left to normalize the result, and the exponent must be decreased by one. When guard digits are present, the digit shifted into the low-order position is 'correct', otherwise it is usually a manufactured zero digit. The intermediate result must also be rounded to fit it back to the working precision of b digits. Of course, rounding is impossible without guard digits. If guard digits are present, several different modes of rounding are possible. The chop mode, or C-mode of rounding simply ignores extra digits, thus truncating the result to working precision. The R-mode of rounding increases the last digit retained whenever the digits being rounded represent more than half an ULP, or unit in the last place retained. Ties can be broken in several ways. Round to nearest even, which forces the last digit retained to be even when rounding occurs, and round to nearest odd are popular choices. Because rounding may cause the intermediate significand to become too large, it may involve an extra normalization shift. Machine designers frequently avoid this remote possibility by rounding before normalizing. Of course, this strategy rounds the wrong digit when renormalization is necessary.

Out-of-bounds results occur whenever the final exponent is too large or too small to be represented in the arithmetic system. Overflow occurs when the exponent is too large. Usually the arithmetic system will signal an error and then continue operation with a default result. This may be the largest representable machine number with an appropriate algebraic sign, or it may be a special reserved operand. Underflow occurs when the exponent is too small. The system may or may not signal an error in this case. Execution usually continues with either the smallest representable number with appropriate algebraic sign, or with a zero result.

We illustrate the effects of underflow on the two calculators that we used earlier by forming the square of $x = 10.0^{-55}$. The result on the TI 59 is the smallest representable positive number, '1.0-99', and a blinking display. This is the result we obtained with the simple algorithm for $|z|$ on that calculator. Evidently, that program stopped when the square of $x=3.0*10^{-50}$ underflowed.

Underflow on the HP 34C results in zero with no indication of error. That explains the result for the simple algorithm for $|z|$ on that calculator. The quantity x^2 underflowed and was replaced by zero. Computation then continued to determine $|z|$ as $\sqrt{(y^2)}$. Because y^2 did not underflow, the program returned '4.00-50', the value of y.

With a little perseverance we can also explain the erroneous result when we used the rectangular to polar conversion program on the TI 59. Suppose that the radial coordinate were determined with the simple algorithm, and that computation continued after underflow with the default result and without the blinking display. Then the program would effectively calculate |z| as $\sqrt{(2.0*10^{-99})}$, which is '4.472136-50' to 7 figures, the result reported by the program.

Curiously, squaring this last result again causes underflow on the TI 59. We now know enough about arithmetic systems to conjecture about what has happened. The system on the TI 59 is a decimal system similar to the systems we discussed above, except that normalized significands lie in the interval [1, 10) instead of the interval [.1, 1). Let X and Y each be $4.472136*10^{-50}$. Then

$$e = e' = -50,$$

and

$$f = f' = 4.472136 .$$

The exponent and significand of the intermediate product are

$$e + e' = -100$$

and

$$f * f' = 20.000 \ldots .$$

At this point the exponent is too small and the significand is not properly normalized. However, renormalization shifts the significand one decimal digit to the right and augments the exponent by 1, thus bringing it back into range. Apparently exponent validity is checked before renormalization. Because the intermediate exponent is out of bounds, underflow is signaled even though the final exponent would be acceptable.

Design errors of this kind are common in contemporary arithmetic systems. They account for most of the strange behavior, such as the vanishing sums and products that we noted above, that plagues implementors of computational software. The failure of simple identities such as

$$1.0 * X = X$$

and

$$X + X = 2.0 * X$$

are all related to the lack of guard digits and/or improperly rounding before renormalization. But these are minor problems compared with the embarrassing failure of the commutative law of multiplication at a level above the rounding threshold on one of our largest and most modern computers. Of course, this was fixed with a design change.

An effort sponsored by the IEEE Computer Society has produced a draft design for an arithmetic system for microprocessors that is almost ideal for computation [1,13,29], and at least one manufacturer is producing prototype chips implementing the complete design. Unfortunately, less than perfect arithmetic systems currently exist and will continue to provide the bulk of our computing power for many years to come. Producers of computational software must be aware of the design flaws in the various computers they use if they are to prepare software with the attributes we advocate in the next section.

4. Software Attributes. We concern ourselves here with three properties of software that are independent of the underlying algorithms. These properties are reliability, robustness and transportability.

The first two properties relate to the performance of the software. Reliability refers to the ability of the program to obtain desired numerical results accurately and efficiently. Of course, accuracy and efficiency depend in part on the algorithm being implemented; some algorithms simply cannot return accurate numerical results or are inherently inefficient. But if the proper algorithm has been selected for implementation, then reliability reflects the quality of the implementation. Reliable computational software successfully handles the problem set defined by the underlying numerical analysis, realizing accuracy over that problem set close to the theoretical prediction for the host environment. Improper appreciation of the environment during implementation may degrade reliability by restricting the problem set, the obtainable accuracy, the efficiency of the program, or some combination of these. For example, the evaluation of a power series will either be inaccurate or inefficient unless the number of terms used is related to the precision of the computer. When asymptotic series are used, both the number of terms and the optimal boundary of the domain for the series are related to machine precision.

Robustness, on the other hand, refers to the ability of the program to avoid or gracefully recover from computational difficulties without unnecessary interruption of program execution. The choice of terminology here is unfortunate because 'robust' has other meanings in related fields, such as statistics. Nevertheless, it is now an accepted software concept. Intuitively, robust software behaves as good software ought to behave. It is resilient and forgiving under misuse, trapping improper arguments and warning the user rather than blindly continuing until it either reaches an impasse or, worse yet, returns incorrect but reasonable looking results with no indication of trouble. Our test data demonstrated that neither the simple program we wrote for the HP 34C nor the rectangular to polar conversion program in the TI 59 calculator are robust because they sometimes return reasonable but incorrect results without warning.

Robust software does not frivolously generate error returns, but when an error return is necessary, it provides precise diagnostic information. Consider the problem of underflow, for example. Usually when underflow occurs an error message is generated and execution continues with a zero result. If the underflow significantly alters the final computed result it is destructive; otherwise it is non-destructive. The trouble with the underflow message is that it does not distinguish between the destructive and non-destructive cases. The user therefore does not know whether it is important or not. Ideally, robust software is free of underflow, so the question of its importance never arises. This requires restructuring the program wherever necessary, consistent with the requirements of reliability, to avoid computations that might lead to underflow. Should underflow be unavoidable, the non-destructive variety is quietly bypassed with an implanted zero result, and the destructive variety is trapped with an appropriate error return, including precise diagnostic information.

The third important attribute is transportability. Although we discuss it here in the context of Fortran programs, the concepts apply to programs written in any algebraic language. One advantage to writing programs in an algebraic language is that they can then be moved to other environments that accept the same source language. Ideally the program would execute correctly in a new environment without change of any kind. We say that the program is portable when that happens. Unfortunately, only the simplest of programs are truly portable in practice. We showed above that the details of a program may have to change to reflect the design of the host arithmetic system. We will see in a moment that details of different language processors may also dictate changes to a program. If a program can be moved and brought up to performance specifications with only a few well documented modifications, we say that the program is transportable instead of portable.

The programming rules to achieve transportability begin with the rules to achieve clarity, i.e., to write programs that are explicit and easily understood (see the excellent monograph [20]). Comments, indentation to highlight control structure, and top to bottom flow with strictly increasing statement labels all contribute to clarity. But transportability requires more than that. Because we cannot be definitive in an article of this length, we suggest the necessary writing style with a few examples.

First, the language constructs must be universally acceptable and understood by language processors. Local enhancements to a language, no matter how alluring, should be avoided. Thus the IMPLICIT statement available in some Fortran dialects should not be used. Similarly, Hollerith strings in formats should be preceded by the appropriate 'nH' specification instead of being set off with special characters such as asterisks or quotes. In a more subtle vein, Hollerith data items should be stored one character per variable because the packing factor varies from one environment to another. It is necessary to restrict the language used to a universally accepted subset of the formal language specified in the standard. For Fortran, the PFORT verifier [24] is a transportable pre-processor that defines a portable subset of ANSI X3.9-1966 Fortran [2] and checks Fortran source code for compliance with that subset. Programs that pass the PFORT verifier can probably be moved about with ease.

Transportability requires more than formal acceptance by a language processor, however. It also requires that the transported program retain its reliability and robustness. We expect that some changes will be required in moving computational software. We would like the software to be written so that necessary changes can be made easily with a text-editor, or perhaps even semi-automatically with a pre-processor of some sort. This implies among other things that every variable be explicitly declared in a TYPE statement, and that real constants contain an explicit E or D exponent. It suggests that all type conversions be explicit using the intrinsic functions such as FLOAT and INT. These and other suggestions are discussed in detail in [25] which we strongly recommend to the reader.

Finally, transportability is enhanced when dependence on the arithmetic environment is made explicit and parameterized. For example, tests to detect the possibility of overflow might use a parameter, XMAX, say, for the largest representable floating-point number in the host environment. The value of XMAX might then be specified in a DATA statement. Then adaptation to a different environment would merely involve changing the value of XMAX in the DATA statement.

At the moment there appear to be two different practical ways to handle these parameters. The Fortran program MACHAR [12] dynamically determines 13 different environmental parameters on most contemporary computers. On the one hand, this program returns realistic values when it works; on the other hand, it is known to fail for certain existing architectures and there is no guarantee that it will work correctly on future architectures. As an alternative, the PORT library contains several machine dependent subroutines that return specific parameters on request [17]. These subroutines contain the parameter values in DATA statements that must be changed as the library is moved. The advantage to this approach is that all such changes for the library are localized in these special subroutines. The disadvantage is that someone must determine the appropriate parameter values for each new environment (values provided in machine manuals are often incorrect, reflecting what the manufacturer thought he did, as opposed to reality). Nevertheless, many newer computational software packages successfully exploit one or the other of these two approaches.

Several recent efforts have attempted to standardize the parameters and related intrinsic functions (see [10] for a discussion of older proposals). The proposal from IFIP WG 2.5 [14] specifying parameter names focused attention on the problem but was not a good solution. The most attractive proposal to date is Brown's approach [5,6] of modeling an arithmetic subsystem of the host environment by specifying conservative values for key environmental parameters. The proposal also includes intrinsic functions for manipulating components of floating-point numbers, such as specifying an exponent or retrieving a significand (similar functions are extensively used in [12] for expository purposes). The ANSI Fortran Standards Committee, X3J3, has tentatively accepted an extensive environmental parameter proposal for inclusion in the next standard [26]. This proposal is based on Brown's work, but extends it somewhat to include one additional intrinsic function for determining 'neighbors' of a floating-point argument.

5. An Example. We illustrate the ideas just presented by designing an algorithm for $|z|$ that will return results usually correct to within rounding error whenever those results can be represented within the arithmetic system being used, and that avoids overflow and underflow. We then describe its implementation in a transportable Fortran program.

Recall that the simple algorithm described earlier for $|z|$ used direct computation of x^2 and y^2 and sometimes resulted in destructive underflow. The following algorithm, drawn from the folklore of early computing, avoids this problem. Using the previous terminology, define

$$u = \max(|x|,|y|)$$

and

$$v = \min(|x|,|y|).$$

Then calculate

$$w = 1.0 + (v/u)^2$$

and

$$|z| = u * \sqrt{w}.$$

Because of the definitions of u and v, $1 \le w \le 2$. Therefore, any underflow that occurs in evaluating w is clearly non-destructive. Overflow is only possible in the final product to form $|z|$, but it is essential in that case because then $|z|$ is too large to be represented in the arithmetic system. Of course, implementations of this algorithm must sidestep explicit overflow and underflow with appropriate tests if they are to be robust. We believe that this algorithm is similar to the algorithm used in the rectangular to polar conversion program in the HP 34C calculator.

This algorithm does not meet our design goals on hexadecimal computers, however, because of wobbling precision. We have $1 \le \sqrt{w} < 2$. In a hexadecimal system

$$\sqrt{w} = 16^1 f,$$

where $1/16 \le f < 1/8$; hence f contains at most t-3 significant bits. This limits the significance of $|z|$ to at most t-3 bits regardless of the normalization

of the significand of $|z|$. An alternative algorithm evaluates $|z|$ as

$$w' = .25 + (v/2u)^2$$

and

$$|z| = 2u\sqrt{(w')},$$

where $1/4 \leq w' \leq 1/2$, and $1/2 \leq \sqrt{(w')} < 1$. The significand of $\sqrt{(w')}$ contains t potentially significant bits and does not limit the significance of $|z|$. The apparent loss of significance because w' contains at most t-1 significant bits is not a real loss. The relative error in the square root of a number is only half of the relative error in the number itself. That factor of 1/2 recovers the extra bit in the square root.

There are values of $|z|$ for which this last algorithm will be less accurate than the first, but statistically we expect that the last algorithm will be significantly more accurate than the first on hexadecimal machines. This conjecture has been verified experimentally [8].

At this point we have designed a computational algorithm that is cognizant of the computational environment. Still, it is the implementation of this algorithm that is important. The Appendix contains a listing of the Fortran source for one possible implementation on a VAX 11/780. The program is designed to be reliable, robust and transportable. We emphasize, however, that it is only intended to illustrate the software principles discussed above; it is not intended to replace any system-provided CABS program.

First, note the format of the source and the extensive use of comments to document the program and to describe the flow of control. These are intended to make the program more understandable and easier to transport. Transportability has also been addressed in several other specific ways. Machine dependencies are concentrated in two constants, XMAX and IOUT, which are described in comments and specified in DATA statements. Real constants not only contain explicit E exponents, but they are also gathered together in DATA statements; there is no need to search the body of the program for constants when converting to a different precision. All conversions between data types are explicit and use the appropriate intrinsic functions. Each intrinsic function used is listed in a comment, thus facilitating the search for them when converting to a different precision. Each variable used is declared in a type statement, further easing

conversion of precision. Finally, the Hollerith strings in the format use the portable 'nH' designator.

The body of the program has been implemented to be completely free of underflow and overflow. The test to avoid division by zero is obvious, but the tests to avoid underflow and overflow are subtle and depend on some knowledge of machine arithmetic systems. The primary test to avoid underflow compares U + V with U. If V is so small in comparison to U that U + V = U to machine precision, then $|z|$ = U to machine precision also, and we need not form V/U. On the other hand, if the sum is different from U in the machine, then the exponents on U and V are close enough that $(V/U)^2$ will not underflow.

Unfortunately, the sum in the test to avoid underflow may overflow if U and V are each large. In that case, the comparison of U with XMAX - V bypasses the underflow test and transfers to the computation of V/U. The quotient will not underflow here because the exponents of U and V are again close to one another.

Note that the expression (V/U)*HALF is used in calculating W in preference to either V/(U+U) or V/(2.0E0*U). This is because the denominators in the latter two expressions could overflow. The explicit test to avoid overflow in the final multiplication of ROOTW by U should again be obvious. The only other subtlety is the sum in the statement labeled 200. This is used in preference to the expression 2.0E0*RES because it is often faster to add than it is to multiply, and the sum is more accurate (sometimes by as much as a decimal place) on machines that lack guard digits or round before renormalization.

This program has been compiled and successfully run on the VAX 11/780. It has also been transported and successfully run on the IBM 3033. The <u>only</u> change made in that move was to respecify XMAX in the DATA statement. Performance on each machine was as expected. Accuracy approached machine precision and compared favorably with the accuracy of the corresponding system programs. Attempts to provoke overflow and underflow all failed. We believe that the software we tested is a faithful representation of the algorithm we designed above, and that it is reliable, robust and transportable.

Do not be misled by this simple example. The production of high-quality computational software is a demanding task. Even generalizing the problem of computing $|z|$ to the outwardly similar problem of computing the Euclidean norm of a vector in n-space for $n > 2$ leads to extremely complicated algorithms, and to even more complicated software implementations. The best method to date for this computation is probably Blue's algorithm [4], and it is almost obscene.

References

[1] ACM SIGNUM Newsletter, Special Issue on the Proposed IEEE Floating-Point Standard, October, 1979.

[2] ANSI, American National Standard FORTRAN, ANSI X3.9-1966, American National Standards Institute, New York, 1966.

[3] ------, American National Standard Programming Language FORTRAN, ANSI X3.9-1978, American National Standards Institute, New York, 1978.

[4] J. L. Blue, "A portable program to find the Euclidean norm of a vector," TOMS, 4 (1978), pp. 15-23.

[5] W. S. Brown, A Simple but Realistic Model of Floating-Point Computation, Computing Science Technical Report No. 83, Bell Laboratories, Murray Hill, N.J., 1980.

[6] ------ and S. I. Feldman, "Environment parameters and basic functions for floating-point computation," TOMS, 6 (1980), pp. 510-523.

[7] W. J. Cody, "The influence of machine design on numerical algorithms," AFIPS Conf. Proc., Vol. 30, 1967 SJCC, Thompson Book Co., Washington, D.C., 1967, pp. 305-309.

[8] ------, "The construction of numerical subroutine libraries," SIAM Review, 16 (1974), pp. 36-46.

[9] ------, "An overview of software development for special functions," Lecture Notes in Mathematics, 506, Numerical Analysis Dundee 1975, G. A. Watson (ed.), Springer Verlag, Berlin, 1976, pp. 38-48.

[10] ------, "Machine parameters for numerical analysis," Lecture Notes in Computer Science, Vol. 57: Portability of Mathematical Software, W. Cowell (ed.), Springer Verlag, New York, 1977, pp. 49-67.

[11] ------, "The challenge in numerical software for minicomputers," Proceedings of the 1st Annual Rocky Mountain Symposium on Microcomputers, IEEE, Inc., New York, 1977, pp. 1-23.

[12] ------ and W. Waite, Software Manual for the Elementary Functions, Prentice Hall, Englewood Cliffs, N.J., 1980.

[13] J. T. Coonen, "An implementation guide to a proposed standard for floating-point arithmetic," Computer, 13, no. 1 (Jan. 1980), pp. 68-79.

[14] B. Ford, "Parameterization of the environment for transportable numerical software," TOMS, 4 (1978), pp. 100-103.

[15] G. E. Forsythe, "What is a satisfactory quadratic equation solver?", Constructive Aspects of the Fundamental Theorem of Algebra, B. Dejon and P. Henrici (eds.), Wiley-Interscience, New York, 1969, pp. 53-61.

[16] ------, "Pitfalls in computation, or why a math book isn't enough," Amer. Math. Monthly, 77 (1970), pp. 931-956.

[17] P. A. Fox, A. D. Hall and N. L. Schryer, "Algorithm 528, framework for a portable library," TOMS, 4 (1978), pp. 177-188. (Algorithm headings only. See Collected Algor. ACM for complete programs.)

[18] W. Kahan, Implementation of Algorithms, Part I., Tech. Report 20, Dept. of Computer Science, University of California, Berkeley, 1973.

[19] ------ and B. N. Parlett, Can you count on your calculator?, Memorandum UCB/ERL M77/21, Electronics Research Lab, University of California, Berkeley, April, 1977.

[20] B. W. Kernighan and P. J. Plauger, The Elements of Programming Style, McGraw-Hill, New York, 1974.

[21] D. E. Knuth, _The Art of Computer Programming, Vol. 2_, Addison Wesley, Reading, Mass., 1969.

[22] H. Kuki and W. J. Cody, "A statistical study of the accuracy of floating-point number systems," _Comm. ACM_, _16_ (1973), pp. 223-230.

[23] A. H. Morris, Jr., _Development of Mathematical Software and Mathematical Software Libraries, Report NSWC TR 79-102_, Naval Surface Weapons Center, Dahlgren, Virginia, 1979.

[24] B. G. Ryder, "The PFORT verifier," _Software Practice and Experience_, _4_ (1974), pp. 359-377.

[25] B. T. Smith, "Fortran poisoning and antidotes," _Lecture Notes in Computer Science, Vol. 57: Portability of Mathematical Software_, W. Cowell (ed.), Springer Verlag, New York, 1977, pp. 178-256.

[26] ------, private communication.

[27] ------, J. M. Boyle and W. J. Cody, "The NATS approach to quality software," _Software for Numerical Mathematics_, D. J. Evans (ed.), Academic Press, New York, 1974, pp. 393-405.

[28] I. A. Stegun and M. Abramowitz, "Pitfalls in computation," _J. Soc. Indust. Appl. Math._, _4_ (1956), pp. 207-219.

[29] D. Stevenson, "A proposed standard for binary floating-point arithmetic," Draft 8.0 of IEEE Task P754, _Computer_, _14_, no. 3 (March 1981), pp. 51-62.

Appendix

```
      REAL FUNCTION CMAG(Z)
C-----------------------------------------------------------
C
C     !*!*!*!*!*!*!*! WARNING !*!*!*!*!*!*!*!*!*!*!
C
C     THIS PROGRAM IS FOR ILLUSTRATIVE PURPOSES ONLY.
C          IT IS NOT A PRODUCTION PROGRAM.
C
C     !*!*!*!*!*!*!*!*!*!*!*!*!*!*!*!*!*!*!*!*!*!*!*!
C
C  THIS FUNCTION TYPE SUBPROGRAM CALCULATES THE MAGNITUDE
C  OF A COMPLEX ARGUMENT Z.
C
C  DATA
C
C     Z - THE COMPLEX ARGUMENT
C
C  RESULT
C
C     CMAG - THE MAGNITUDE OF Z WHEN THAT QUANTITY IS REPRESENTABLE,
C            AND XMAX, THE LARGEST REPRESENTABLE NUMBER, OTHERWISE.
C
C  ERROR RETURNS
C
C     WHEN THE MAGNITUDE OF Z IS NOT REPRESENTABLE, THE SUBPROGRAM
C     WRITES AN ERROR MESSAGE TO THE STANDARD OUTPUT UNIT AND
C     RETURNS THE LARGEST REPRESENTABLE FLOATING-POINT NUMBER.
C
C  MACHINE DEPENDENCIES
C
C     XMAX - AN ESTIMATE OF THE LARGEST REPRESENTABLE FLOATING-POINT
C            NUMBER WHICH CAN BE USED IN ARITHMETIC OPERATIONS MUST
C            BE SPECIFIED IN THE DATA STATEMENT.
C     IOUT - THE LOGICAL DESIGNATION OF THE STANDARD OUTPUT UNIT
C            MUST BE SPECIFIED IN THE DATA STATEMENT.
```

```
C
C  INTRINSIC FUNCTIONS USED
C
C     ABS, AIMAG, AMAX1, AMIN1, REAL, SQRT
C
C
C  LATEST REVISION - SEPTEMBER 3, 1980
C
C  AUTHOR - W. J. CODY
C             ARGONNE NATIONAL LABORATORY
C
C------------------------------------------------------------------
      COMPLEX Z
      REAL FOURTH,HALF,RES,ROOTW,U,V,W,X,XMAX,Y,ZERO
      INTEGER IOUT
      DATA FOURTH/0.25E0/,HALF/0.5E0/,ZERO/0.0E0/
      DATA IOUT/6/,XMAX/0.17E39/
C------------------------------------------------------------------
      X = ABS(REAL(Z))
      Y = ABS(AIMAG(Z))
      U = AMAX1(X,Y)
      V = AMIN1(X,Y)
C------------------------------------------------------------------
C  TEST TO AVOID DIVISION BY ZERO
C------------------------------------------------------------------
      IF (U .EQ. ZERO) GO TO 150
C------------------------------------------------------------------
C  TEST TO AVOID UNDERFLOW
C------------------------------------------------------------------
      W = ZERO
      IF (U .GT. XMAX-V) GO TO 40
      IF (U+V .EQ. U) GO TO 50
   40 W = (V/U) * HALF
   50 ROOTW = SQRT(FOURTH+W*W)
C------------------------------------------------------------------
C  TEST TO AVOID OVERFLOW
C------------------------------------------------------------------
      RES = XMAX * HALF
```

```
      IF (RES/ROOTW .LE. U) GO TO 100
      RES = U * ROOTW
      GO TO 200
C-----------------------------------------------------------
C  WRITE OVERFLOW ERROR MESSAGE
C-----------------------------------------------------------
  100 WRITE (IOUT,9999)
      GO TO 200
C-----------------------------------------------------------
C  RETURN
C-----------------------------------------------------------
  150 RES = ZERO
  200 CMAG = RES + RES
      RETURN
 9999 FORMAT(38H0ERROR IN CMAG, RESULT WOULD OVERFLOW. /
     1       41H COMPUTATION CONTINUES WITH RESULT = XMAX //)
C  ---------- LAST CARD OF CMAG ----------
      END
```

Implementation and Testing of Function Software

W. J. Cody*

Applied Mathematics Division
Argonne National Laboratory
Argonne, Illinois 60439

1. Introduction. In some respects the implementation and testing of software for the evaluation of elementary and special functions is much simpler than the implementation and testing of software for other computational purposes. Certainly the methods for evaluating functions are better understood than methods for, say, minimizing a function of several variables. The evaluation of a function is generally a simple one-to-one mapping from a one or two dimensional domain to a one or two dimensional range, and the concept of accuracy in that mapping is well understood. Function minimization, on the other hand, is a complicated mapping from one high dimensional space to another with a poorly understood concept of accuracy or measure of success for the process. However, because function evaluation is relatively easy we expect the software to return precise results. This implies that function software is more sensitive to the computational environment than software for function minimization, and that testing procedures for function software must be more carefully designed and implemented than those for function minimization software. This paper is an overview of proven techniques for preparing and testing function software. It is not a survey of algorithmic methods for function evaluation; the best work of that sort is [8].

---

* This work was supported by the Applied Mathematical Sciences Research Program (KC-04-02) of the Office of Energy Research of the U.S. Department of Energy under Contract W-31-109-Eng-38.

We assume the material in [6] as background. The next section summarizes some of the design problems peculiar to software for the elementary functions. Section 3 enlarges on these ideas and illustrates them with an outline of an algorithm for the evaluation of the gamma function for a real argument. Section 4 treats testing objectives in general and discusses the details of one widely used but machine sensitive technique for determining the accuracy of function software. Section 5 discusses a less sensitive but transportable set of Fortran programs, the ELEFUNT package, for performance evaluation of elementary function software.

Much of the material presented here has appeared before (see, in particular, previous publications cited in the bibliography). Many individuals have contributed to our ideas without necessarily endorsing them. We cannot individually acknowledge all who have influenced our thinking, but we must single out W. Kahan, the late H. Kuki, and our colleagues at Argonne National Laboratory. We express our gratitude to an anonymous multitude of others.

2. Elementary Functions. Software for the elementary functions normally resides in system libraries accompanying compilers for high level languages. Unless there is strong evidence of poor performance, users tend to regard these programs in the same way they regard the arithmetic operations in the computer. That is, they view them as friendly 'black boxes' that can be trusted to be efficient and accurate. Only careful preparation of software guarantees that the trust will not be violated.

Of the three characteristics of good software discussed in [6], transportability is the least important in this case. Elementary functions as a class are undoubtedly the least transportable of all computational software. Accuracy and efficiency considerations usually dictate that the programs be written in assembly language. They often exploit special machine instructions unique to the local environment and not generally available through higher level languages. Thus algorithms and implementations are highly customized and transportability is moot.

Despite this customization, good elementary function programs share certain design principles. In general the computation of an elementary function involves three steps. First, the given argument is usually reduced to a related argument in some restricted domain together with parameters describing the argument reduction. For example, in computing the sine function for radian arguments a given argument x might be decomposed into

$$x = |x| \text{ sign}(x)$$

and

$$|x| = N\pi + g,$$

where $|g| \leq \pi/2$ and N is an integer. Here g is called the <u>reduced argument</u>. In the second step, a related function value is calculated for the reduced argument. This might involve the evaluation of a polynomial or rational approximation of some sort, or it might involve a more complex evaluation such as in the CORDIC scheme [14]. Finally, the desired function value must be reconstructed from the results of the first two steps. Thus

$$\text{sine}(x) = (-1)^N \text{sign}(x) \text{ sine}(g).$$

All the parameters derived in the argument reduction step have been used here, but that is not always so.

The accuracy of an elementary function program depends on the accuracy of the reduced argument as well as the accuracy of the approximation. Because of its apparent simplicity, argument reduction is often slighted during implementation with most of the effort going into the approximation. The result is that many elementary function programs cannot determine accurate function values for larger arguments. Table I compares the computation of sine(22) using the built-in functions in a variety of hand calculators with a standard value taken from [1]. (All calculator results are quoted to the working precision of the particular calculator.) Additional table entries give results for some calculators when careful argument reduction is used before calling the built-in function, and there is also an entry for a completely new program on the TI 59. Details of the argument reduction scheme and the new program are developed below. The computation of sine(22) is a severe test of the argument reduction step because the argument is close to an integral multiple of π. The built-in program on the HP 34C returns a correct result within a unit in the last place (an ULP), while the results for the other built-in programs are grossly incorrect in the last few digits. The error is correctable on the HP 65 with some preliminary argument reduction, suggesting that the built-in argument

Table I

Accuracy of sine(22) on selected calculators

Program	sine(22)	
Standard Value [1]	-8.8513 09290 40388	E-03
TI 59 (hardware)	-8.8513 09285 516	E-03
TI 59 (careful arg. red.)	-8.8513 09289 517	E-03
TI 59 (new program)	-8.8513 09290 405	E-03
HP 65 (hardware)	-8.8513 06326	E-03
HP 65 (careful arg. red.)	-8.8513 09290	E-03
HP 34C (hardware)	-8.8513 09289	E-03

reduction is faulty. Careful preliminary argument reduction does not help on the TI 59, suggesting a more serious problem. We cannot draw any conclusions about the accuracy of the approximation being used, however, because the sample is too small.

To understand the problems in argument reduction we must first understand where errors originate. There are two types of error associated with computational software in general. The first is caused by error in the data. Let $y = f(x)$ be a differentiable function of x. Then

$$dy/y = x\, f'(x)/f(x) \quad dx/x$$

is an analytic relation between the relative error dx/x in the argument and the relative error dy/y in the function value. The <u>transmitted error</u> dy/y depends solely on the <u>inherited error</u> dx/x and the analytic properties of the function. It does not involve computations internal to the software, and therefore is beyond the control of the software. All error directly attributable to the internal computations, such as rounding, truncation of analytic expansions, and inexact representation of constants is grouped together and called <u>generated error</u>. This is the error that is the sole responsibility of the computational software. Because the software cannot detect inherited error, it should treat all data as if it were exact.

The determination of the reduced argument

$$g = |x| - N\pi$$

involves the difference of two nearly equal quantities. By assumption $|x|$ is exact, but $N\pi$ cannot be represented exactly in any finite precision and must involve some rounding error. As $|x|$ becomes larger more and more leading digits of $|x|$ are lost in the subtraction, thus promoting the rounding error in $N\pi$ in importance. The relative error in the reduced argument g is therefore roughly proportional to $|x|$. This error can be controlled if the product $N\pi$ and the difference are both carried out in higher precision arithmetic. The extension of $|x|$ to higher precision does not introduce error because $|x|$ is assumed to be exact. π is a mathematical constant known to within rounding error in any precision. The only problem, then, is in the arithmetic operations. When higher precision arithmetic is available in hardware it should be used. When it is not available in hardware the following computation in working precision effectively extends the precision of the operations. Evaluate

$$g = (|x| - N*C) - N*C',$$

where C and C' are working precision constants with the property that $C+C' = \pi$ to beyond working precision. Thus C is a machine number containing the first few base-B digits of π, where B is the radix of the floating-point system, plus a number of trailing zero digits, and C' is a full precision representation of π-C.

Using this two-step process, the computed value of g is correct to within a few ULPs when the product N*C can be represented exactly in the active arithmetic register, i.e., if N is not too large. When this condition is not met the accuracy of the process reduces to that of the one-step scheme. This scheme protects the precision of g only to the extent that the representation of π by C+C' exceeds working precision. If $|x|$ and $N\pi$ should agree to more significant digits than are contained in C, then the low order digits of g will again probably be in error. Other subtleties that we have ignored here are discussed in [7].

The effectiveness of this two-step argument reduction is demonstrated by the results for the HP 65 programs in Table I. Here, careful argument reduction retains full precision in the result even though four leading decimals are lost in the subtraction.

The restriction on the magnitude of N, hence on the magnitude of x, imposed by this scheme is not as bad as it might appear. Consider the case where $|x|$ is so large that all its significant digits correspond to the integer part of $N\pi$. Then it is clearly unreasonable to attach any meaning to g. If g retains full significance for small $|x|$, and no significance for large $|x|$, then there must be some point where the user should be warned through an error return that the result of the computation cannot be guaranteed to be accurate. The point where N*C cannot be represented exactly in the arithmetic register is a convenient and reasonable threshold for such an error return.

The argument reduction step is often the best place for two related programs to be merged. The sine program can serve double duty, for example, by being used for the cosine as well. Because

$$\text{cosine}(x) = \text{sine}(x+\pi/2)$$

and

$$\text{cosine}(-x) = \text{cosine}(x),$$

cosine(x) can be calculated by evaluating sine($|x|+\pi/2$). But this computation must be done carefully. Simple addition of $\pi/2$ to $|x|$ introduces argument error that defeats subsequent careful argument reduction. Full precision is maintained in g if, instead, the value of N is adjusted during argument reduction.

Retaining full precision in the second step, the evaluation of the function approximation, is ordinarily an easy task. The approximation used is often a simple rational function that retains much of the analytic behavior of the function being approximated. In the sine function, for example, the approximation

$$\text{sine}(x) = x + x\,R(x^2),$$

where R is either a rational or a polynomial function, preserves the properties

$$\text{sine}(-x) = -\text{sine}(x)$$

and

$$\text{sine}(x) \sim x, \qquad |x| \ll 1.$$

The primary concerns in this step are the buildup of roundoff error and the loss of significance through subtraction. Polynomial forms are most efficiently evaluated using nested multiplication. When that computation is numerically unstable, conversion to an equivalent Chebyshev polynomial form coupled with Clenshaw's evaluation scheme [2,13] or to a minimal Newton form [11] is often completely stable although less efficient. Rational functions are more versatile than polynomials and can be modified in many ways to improve stability. Not only can the numerator and denominator be modified individually as polynomials, but various special representations such as continued fraction type expansions are also possible. The rational form suggested above for the sine function is an example where the strong numerical dominance of a single term has been made explicit. In this particular case numerical stability is achieved by casting the approximation as a dominant term plus a correction term.

The remaining concern at this point is the avoidance of underflow and overflow. Usually underflow here is non-destructive and the primary motive is to avoid issuing a misleading error indicator. For the sine function, for example, the formation of x^2 will underflow when $|x|$ is small enough. When that happens, x represents the function value to machine precision. No special action is necessary if the arithmetic system quietly replaces underflow with zero. In all other cases the formation of x^2 must be bypassed when $|x| < \varepsilon$ for some machine dependent ε, and the function value x must be returned.

The algorithm just outlined is given in greater detail in [7] and is the basis for the new program on the TI 59 identified in Table I. We will examine its performance on the TI 59 in greater detail later.

The care taken in implementing the elementary functions does not guarantee that all computations involving those functions will be correct to within a few ULPs. What is guaranteed is that if the argument is exact and within certain bounds, then the results of the function evaluation itself will be correct to within a few ULPs. This guarantee is similar to that implied by the floating-point arithmetic system. Multiplication, for example, is assumed to be correct to within 'rounding' if the result stays within certain bounds, but the result is 'correct' only if the operands are.

Even the result obtained by following one elementary function by another cannot be guaranteed within a few ULPs. Consider, for example, the task of evaluating $z = x^y$. The mathematical definition of z is

$$z = \exp[y \ln(x)].$$

If x and y are exact, it seems reasonable to expect that z should be correct to within a few ULPs if the exponential and logarithm routines are. But this is not necessarily the case. A little analysis shows why. Let

$$w = y \ln(x)$$

and

$$z = \exp(w).$$

Then the relative error in z equals the absolute error in w. That is,

$$dz/z = dw.$$

The absolute error in w is proportional to the magnitude of w because of the finite machine precision. Thus the limiting factor is the finite word length and not the accuracy of the exponential and logarithm functions. The correct way to evaluate x^y is to retain extra precision in all intermediate results. This requires a special self-contained program [7] independent of the existing logarithm and exponential programs. Table II compares the results of calculating 2.5^{125} using first the logarithm and exponential functions and then the built-in power function on the TI 59 and HP 34C. The incorrect results on the TI 59 agree because the built-in power function uses the existing logarithm and exponential functions, neither of which is necessarily correct to within rounding error. The combination of logarithm and exponential on the HP 34C results in a large error, as predicted above, despite the individual functions on that machine being correct to within rounding error.

The phenomenon we have just demonstrated is general and must be clearly understood. Results obtained from the composition of two or more items of software each reliable in themselves may not be accurate. Reliability in numerical software only assures the user that the software properly processes the data at hand. When large errors are detected in a computation, this assurance then suggests that the trouble lies in problem formulation, programming errors,

Table II

Accuracy of 2.5 ** 125 on selected calculators

Program	2.5 ** 125	
Standard Value [1]	5.5271 47875 26044	E 49
TI 59 (exp(ln))	5.5271 47875 888	E 49
TI 59 (power)	5.5271 47875 888	E 49
HP 34C (exp(ln))	5.5271 47962	E 49
HP 34C (power)	5.5271 47875	E 49

or composition of software, rather than in the performance of the individual software items.

3. Special Functions. The evaluation of special functions of mathematical physics ordinarily does not involve argument reduction. Instead, the function domain is usually divided into several sub-domains and appropriately tailored algorithms are implemented over each. The large variety of possible types of algorithms is illustrated in Gautschi's survey [8]. Some functions are best evaluated using recurrence, others from analytic expansions, and still others using subtle methods such as the Landen transform. In every case, however, reliability and robustness depend on the care taken in selecting and implementing the algorithm.

To illustrate these ideas, the following discussion of the design of a program to evaluate the function $\Gamma(x)$ for real arguments is adapted from [5]. Under the assumption that the argument is exact, the program is to produce an accurate function value whenever that value exists, can be approximated in the machine and can be obtained with reasonable effort. In all other cases, an error exit is to be taken. The computation is also to be free of overflow and underflow.

For $x > 0$ the recurrence relation

$$\Gamma(x+1) = x\,\Gamma(x)$$

can be used to reduce the computation to that for $\Gamma(x)$ over some interval of unit length. Examination of a plot of the function suggests that either of the intervals [1.0,2.0] or [2.0,3.0] be used. Because $0.5 < \Gamma(x) \leq 1$ for x in the first interval, the hexadecimal representation of the function value is guaranteed to contain the maximum number of potentially significant bits [6]. On the other hand, $1 \leq \Gamma(x) \leq 2$ for arguments in the second interval, and the hexadecimal representation of the function value contains three fewer potentially significant bits. Because the significance of any function value computed from the recurrence depends on the significance of the value in this subdomain, the first interval should be used whenever there is any possibility that the program will be executed on a hexadecimal machine. We would normally choose that interval just to be safe, although rational or Chebyshev polynomial expansions are available in the literature for either interval [2,10].

Repeated use of the recurrence relation is inefficient when x becomes large. Instead, we evaluate

$$\Gamma(x) = \exp[\ln \Gamma(X)]$$

directly using either the asymptotic expansion or a rational approximation such as those in [10] for the evaluation of $\ln \Gamma(x)$. Either way, this computation involves a composition of functions that appears to be subject to the same error growth we discussed for the power function. Both the asymptotic form and the rational approximation contain a dominant term $(x-1/2)\ \ln(x)$, however, that becomes a factor of $x^{(x-1/2)}$ for the gamma function. The accuracy of the computation can be improved if this term is evaluated separately from the rest of the approximation and if the power function is reliable.

The gamma function is monotonic increasing for increasing positive x. Thus an error return must be provided for x > XBIG, where

$$\Gamma(\mathrm{XBIG}) = \mathrm{XMAX},$$

and XMAX is the largest machine representable number. The correct value for

XBIG in any particular environment is easily determined with Newton's method and an initial guess obtained from the standard asymptotic expressions for $\Gamma(x)$ and $\Gamma'(x)/\Gamma(x)$.

The recurrence relation shows that $\Gamma(x) \to 1/x$ as $|x| \to 0$, hence an error return must also be provided for $|x| <$ XMININV, where XMININV is the smallest representable positive number whose reciprocal is also representable. If XMININV = XMIN, the smallest representable positive number, then $\Gamma(x)$ is computable for all small non-zero x and no special error returns are necessary.

There is a small region, XMININV $\le x <$ XSMALL, for which

$$\Gamma(x) = 1/x$$

to machine precision, and the more complicated evaluation of $\Gamma(x)$ can be bypassed. Let

$$\varepsilon = B^{(-b-1)},$$

where there are b base-B digits in the significand of a floating-point number. Then

$$1 + \varepsilon = 1$$

on the machine,

$$\Gamma(1+\varepsilon) = \Gamma(1) = 1,$$

and XSMALL = ε.

The reflection formula

$$\Gamma(x) = \pi/[\text{sine}(\pi x)\ \Gamma(1-x)], \quad x < 0,$$

reduces the computation for negative arguments to one for positive arguments. The evaluation of sine(πx) is the key here, because rounding error in evaluating πx appears as inherited error to the sine routine. However, if $x = -n+f$, where f is a fraction, then

$$\text{sine}(\pi x) = (-1)^{-n}\text{sine}(\pi f).$$

Some preliminary argument reduction at this point minimizes the inherited error for the sine function.

Preliminary argument reduction also provides the opportunity for an easy test for singularities. $\Gamma(x)$ does not exist when $x = -n$, i.e., when $f = 0$. Mathematically, this is the only case where $\Gamma(x)$ does not exist, but computationally it is possible that $\Gamma(-n+\delta)$ will overflow for sufficiently small δ. The case $n = 0$ has already been treated, so assume $n > 0$. Now, n cannot be too large or $1/\Gamma(1-x)$ will underflow in the reflection formula. Thus $|\delta| \geq |\varepsilon n|$ and $\text{sine}(\pi\delta) \geq n\pi\varepsilon$ in the machine. From the reflection formula

$$\pi/\Gamma(1-x) \leq \Gamma(x) < B^b/[n\ \Gamma(1-x)] < 2B^b/n.$$

This shows that $|\Gamma(f)|$ will not overflow for $f \neq 0$.

There is still a lower bound for x below which $\Gamma(x)$ may underflow. The determination and use of a precise lower bound are difficult, probably falling in the design category of requiring excessive effort. The problem is that $\Gamma(x)$ is representable for $x = -n+f$ and small $|f|$, but may underflow when $f = 1/2$. An inelegant solution is to restrict x to

$$-x < \min(XNEG, XBIG-1),$$

where XNEG satisfies the equation

$$1/|\Gamma(1+XNEG)| = XMIN.$$

The design just outlined includes several clearly defined machine dependencies. A Fortran program based on this design and properly using the parameters XBIG, XMAX, XMIN, XMININV, XSMALL, ε and XNEG can be modified for a different environment by reinitializing these parameters. The extension of this program design to include the computation of $\ln \Gamma(x)$ is easy, and is left as an exercise for the interested reader.

4. Testing Procedures. There are two fundamentally different reasons for testing computational software, hence two fundamentally different approaches to the task. Creators of new algorithms are intent on showing that their creations are superior in some sense to other algorithms, and they approach performance testing as a contest. Tests are specifically designed to display whatever superiority the new algorithm may have. There is ordinarily no attempt to uncover or explore weaknesses in the algorithm or its implementation.

Someone selecting programs for inclusion in a library is interested in overall performance, however. If some duplication of purpose is acceptable to the library, the concern may be more with eliminating programs that are unacceptable and in matching programs with problem characteristics than with determining the 'best' program. Tests for this purpose ought to aggressively exercise a program in ways that will detect weaknesses as well as show strengths, that will explore robustness as well as problem solving ability. Such testing is analogous to a 'physical examination'. Inevitably the results of such testing will be used to compare programs, but the original intent in this approach is that a program will stand or fall on its own merits.

Tests in the sense of contests are too problem dependent to be applied in general. We therefore limit our discussion here to the second type of testing as it applies to function software. If there is a theme in what follows, it is that testing is an important numerical problem. As much attention should be given to designing and implementing objective test programs as is given to designing and implementing software for solving other numerical problems. Each test performed ought to have a specific purpose just as each step in the solution of any other numerical problem should be purposeful. Blind testing is no more useful than attempting to solve a polynomial equation by trial and error.

A complete battery of tests should include special tests to verify robustness, check error returns, and verify simple analytic properties of the function, along with the usual timing and accuracy tests. Special tests are mostly a matter of ingenuity. For example, because robust programs have been defined as resilient under misuse, tests for robustness should include deliberate misuse, such as specifying arguments violating stated constraints for the program and arguments leading to possible underflow or overflow. Consider the standard Fortran function ATAN2(U,V), for example. This function is undefined when U=V=0, and is defined as arctan(U/V) otherwise. Comprehensive tests of the program should include computations in which all possible combinations of zero and the largest and smallest nonzero machine numbers are used as arguments. If an implementation of ATAN2 is subject to underflow and overflow, one of these argument pairs should discover it. Prudence dictates that the argument pair (0,0) be used last, however, because it should lead to an error that might stop program execution.

Symmetries and other simple properties of a function should also be verified computationally. For example, the properties

$$\exp(-x) = 1/\exp(x)$$

and

$$\ln(1/x) = -\ln(x)$$

could be verified to hold approximately by using just a few arguments (the second relation should hold exactly if both x and 1/x are exactly representable in the machine). More complicated properties, such as monotonicity and periodicity, especially with an irrational period, are too difficult to verify computationally and are probably best ignored during testing.

The details of timing a program depend critically on the computational environment. All we can do here is offer general suggestions and guidelines. One widely used technique is to call the program with fixed arguments inside a loop, calling the clock routine immediately before and after the loop. If the program is short enough that the overhead for the loop is significant, the elapsed time can be adjusted by also timing an empty loop. Care must be taken to make each timing loop long enough to compensate for the inherent coarseness of the clock, and to vary the fixed arguments used to insure that each major path through the program is independently timed. If there is reason to suspect that a program may recognize that the argument has not changed from the last call and may then take a computational shortcut, the loop must include two calls with different arguments that follow the same computational path. Finally, timing may be load dependent, especially under large and complicated operating systems. Under such conditions it is best to repeat the timing runs several times under varying system loads and to average the reported times after discarding any obvious outliers. When several timing runs are averaged, the standard sample deviation adds useful information to the reported average time.

That leaves only the sticky problem of determining the accuracy. We are fortunate, in dealing with function programs, in having a well defined concept of accuracy. A function maps one or more arguments into one real or complex value. The accuracy of the computer program is determined by measuring the difference between the correct function value and the computed function value.

We explore the mechanics of the measuring process here and delay discussing the art of accuracy testing until the next section.

Earlier we identified two types of error associated with numerical software: transmitted error and generated error. Ideally, accuracy tests should measure only the error directly attributable to the software, i.e., the generated error. The first requirement, then, is that the test arguments be exact. If they are not exact, transmitted error will creep into the measurement and distort the findings. This requirement immediately eliminates an obvious testing scheme -- comparison against published tables.

Consider what happens when table comparison is used. First, an argument must be selected from a sparse set and fed into the computer. The argument is ordinarily a decimal fraction, containing only a few significant decimal digits, that cannot be represented exactly in non-decimal number systems. This error associated with the incompatibility of the number systems is compounded by the rounding error generated in the computer I/O conversion program. Thus the argument finally passed to the function program is probably different from the argument in the table. Even if somehow the argument should be exact, the whole process must be repeated to output the computed function value. There is again an I/O conversion subject to rounding error, and the final representation of the function value in decimal form is incompatible with the internal machine representation. With all these sources of contamination, the final comparison between the 'computed' function value and the tabular value tells little about the accuracy of the computer program unless there is huge error. We therefore reject comparison against published tables as a viable means of accuracy testing.

Comparison is still possible, but the comparison should be against standard function values carefully generated within the computer. The technique that we have found most useful and that we now often see being used by others is a direct comparison against higher precision computations for random arguments [3]. Briefly, the procedure is as follows.

First, the desired test interval is subdivided into n subintervals. Next, a random argument is generated in each subinterval using the same precision arithmetic as is used in the function being tested. These arguments are regarded as exact, eliminating inherited error from the testing process. Each argument in turn is then extended to higher precision without changing its numerical value by appending appropriate trailing zero digits to its machine representation. The higher precision argument is then used in a higher precision function program, and the result is taken as the standard function value for the original argument. Next, the original working precision argument is used in the program under test and the resulting function value is compared against the standard value in one of several possible ways. Either the working precision result can be extended to higher precision and the error measured there, or the higher precision result can be rounded back to working precision and the error measured there. The second approach is ordinarily used when standard arguments and function values are precomputed and saved on magnetic tape or other storage devices. Stored values must not pass through I/O conversion, however. They must be stored and retrieved in machine representation, i.e., in binary or hexadecimal form, to preserve their purity.

Statistics gathered from such tests normally include the maximum relative error, MRE, and the root mean square relative error, RMS, where

$$MRE = \max |RE_i|,$$

$$RMS = \sqrt{[1/n \; \mathrm{sum}(RE_i)^2]},$$

$$RE_i = (F_i - f_i) / f_i,$$

n is the number of random arguments x_i used, F_i is the calculated function value and f_i is the higher precision value. If the test arguments have been sorted algebraically, the error can be monitored to detect trends pointing to correctable problems in the function being tested.

A different set of statistics is often collected at the same time as the MRE and RMS. It is possible through programming subterfuge to treat the floating-point function values as large integers and to gather statistics on the ULP errors by taking integer differences. A tabulation of the frequency of ULP errors can be informative. Table III contains test results of the types just described for a good implementation of the sine function on a VAX 11/780.

We have glossed over several problems here that should be explored further. First, we have assumed that the higher precision program is accurate. This

Table III

Test Results for SIN on VAX 11/780

Argument Range	Frequency of Error in ULPs					MRE	RMS
	0	1	2	3	>3		
$(0,\pi/2)$	1936	64	0	0	0	6.61 E-08	2.27 E-08
$(2\pi, 8\pi)$	1928	70	1	0	1	2.53 E-07	2.45 E-08

implies that it too must undergo testing. Fortunately, this requirement does not lead to an infinite loop, because the accuracy required of the higher precision routine is not that great. It need only be accurate to slightly more than the test precision; accuracy to a few ULPs in the higher precision is not needed. Such moderate accuracy can be verified, for example, by simply comparing two higher precision programs implementing different algorithms.

There is also a problem when higher precision arithmetic is not available in the system. In that case either programmed arithmetic must be used, or the standard arguments and function values must be generated in a different system and transmitted to the working system via magnetic tape. This latter approach requires careful control to insure that the arguments and corresponding function values are exactly representable in the target machine even when that representation is not native to the machine generating the values. This is the method that was successfully used in the FUNPACK project [4].

We must not minimize the problems just discussed; their satisfactory solution requires some effort. Preparing tests generally takes much more effort than preparing the software being tested. However, many of the testing procedures just described can be programmed once and for all in transportable subroutines. We now have a collection of Fortran subroutines, for example, that take as arguments the name of the single precision function program to be tested, the name of the double precision function program to be used as a standard, the endpoints of the test intervals, and the number of random arguments to be used. Printed output from these subroutines includes the MRE and RMS errors, and a frequency table of the errors measured in ULPs. The differences between the various subroutines lie in the number and type of arguments expected by the function programs, i.e., one or two, real or complex. It is therefore only necessary to select the proper test subroutine, to provide a driver specifying the functions to be tested and the test intervals, and to provide the master function program when preparing a new set of accuracy tests. Because floating-point numbers are represented differently in different systems, the details of the computation of errors in ULPs must sometimes be modified when transporting the tests. The least significant bit of the significand of a single precision number occurs in the middle of its machine representation on a VAX, for example, instead of at the end as on most other machines, and the subroutine used to generate the data in Table III is different in detail from the corresponding subroutine used on IBM or CDC equipment. Other than that detail, the test programs are reasonably transportable to systems supporting both single and double precision arithmetic. They are not useful, however, for testing the accuracy of double precision programs because few machines support arithmetic of more than double precision.

5. ELEFUNT The discussion in the last section omitted the possibility of determining accuracy by examining the degree to which functions satisfy mathematical identities. This approach to accuracy testing is not new [9,12], but it has not been particularly successful in the past. The main problem is that identities introduce error into the computation that becomes confused with the error generated in the function program. Often the new error completely masks the error that we really want to measure. Nevertheless, recent advances have shown that measuring error in carefully selected identities, implemented with the same care taken in implementing elementary functions, provides useful error statistics.

The motivation for the work leading to this discovery was the development of a collection of self-contained transportable test programs for the elementary functions, the ELEFUNT package, intended to complement a comprehensive collection of algorithms and implementation notes for the computation of these functions [7]. The methods described for accuracy testing in the last section were not suitable for this package because they assumed the availability of higher precision arithmetic and of a suitable master function routine. The only alternative was to resort to the use of identities. We illustrate the approach taken by examining the program to test the sine and cosine functions.

The first task is to select an identity that is not likely to be used internally in the program being tested. Our purpose is to measure the accuracy of the calculation of a particular function and not the accuracy of the implementation of an identity. Identities based on the triple angle formulas are likely candidates for the sine and cosine functions. Thus, we propose to measure

$$E = \{\mathrm{sine}(x) - \mathrm{sine}(x/3)[3 - 4\ \mathrm{sine}^2(x/3)]\} / \mathrm{sine}(x).$$

It is important that the identity not introduce unnecessary error into the computation. Here, proper selection of arguments eliminates the potential loss of significance in the subtraction inside the brackets. When x is drawn from the interval $[3m\pi,(3m+1/2)\pi]$ for integer m, x/3 lies in the interval $[m\pi,(m+1/6)\pi]$, $4\ \mathrm{sine}^2(x/3) \le 1$, and there is no cancellation of leading significant digits. We henceforth assume these limitations on the test arguments.

We stated earlier that testing should be purposeful. There are two distinct computational steps in the evaluation of the sine function that could contribute to the generated error in a computer routine: the argument reduction step and the evaluation of the function for the reduced argument. The accuracy of these steps should be checked separately if at all possible. Restricting the argument to the interval selected by setting m = 0 assures that there will be little, if any, argument reduction involved, and the accuracy statistics gathered will reflect the accuracy of the function evaluation given a reduced argument. Arguments selected from the interval with m = 2 will assure that argument reduction is involved. Significant differences between the results of these tests must then be attributable to the argument reduction scheme.

Results of these tests applied to a double precision sine function on the IBM 195 are given in Table IV under the heading 'Simple Tests'. Master test results for the same function, obtained by comparison against higher precision computations, are also listed in the table. Each test used 2000 random arguments from each interval. Errors are tabulated as the reported loss of base-B digits, where B is the radix for the arithmetic system in the machine. Tests labeled '1' correspond to m = 0, and those labeled '2' correspond to m = 2.

Table IV

Accuracy Tests for Sine/Cosine

Test	Machine	B	Library or Program	Reported Loss of Base B Digits in MRE	RMS
Master Values					
1	IBM 195	16	Argonne	1.20	0.48
2	IBM 195	16	Argonne	1.20	0.56
Simple Tests					
1	IBM 195	16	Argonne	1.28	0.78
2	IBM 195	16	Argonne	3.50	2.30
Tests with Purified Arguments					
1	IBM 195	16	Argonne	1.18	0.69
	PDP/11	2	DOS 8.02	1.99	0.10
	Varian 72	2	Fort E3	1.87	0.00
	TI 59	10	Built-in	2.34	1.66
	TI 59	10	New	1.92	1.30
2	IBM 195	16	Argonne	1.16	0.70
	PDP/11	2	DOS 8.02	1.74	0.09
	Varian 72	2	Fort E3	13.54	8.55
	TI 59	10	Built-in	3.81	2.96
	TI 59	10	New	1.71	1.11
3	IBM 195	16	Argonne	1.16	0.69
	PDP/11	2	DOS 8.02	12.63	8.55
	Varian 72	2	Fort E3	12.69	7.31

These results show that the identity tests we have described are not reliable, especially for the second interval. The test procedure has introduced large errors into the computation completely masking the error generated in the sine routine. To see how, assume for the moment that $|x| \le \pi/2$. The random argument x is exactly representable in the machine, but there may be a small error ε in the evaluation of the argument x/3. It is easily shown that for $|\varepsilon| \ll 1$,

$$\text{sine}(x/3+\varepsilon) = \text{sine}(x/3) + \varepsilon$$

to terms of first order. Further assume that relative errors of D and d are made in the evaluation of sine(x) and of sine(x/3), respectively. Substituting these expressions in the expression for E gives

$$E = \{\text{sine}(x)(1+D) - \text{sine}(x/3+\varepsilon)(1+d)[3 - 4\ \text{sine}^2(x/3-\varepsilon)(1+d)^2]\} \,/\, \{\text{sine}(x)(1+D)\}.$$

Using the original identity and the above relation for sine(x/3+ε), and keeping only terms linear in D, d and ε, this simplifies to

$$E = D - d[1 - 8\ \text{sine}^3(x/3)/\text{sine}(x)] + \varepsilon[1/\text{sine}(x/3) - 8\ \text{sine}^2(x/3)/\text{sine}(x)].$$

Because $\text{sine}^3(x/3)/\text{sine}(x)$ is bounded above by 1/8 for the interval under consideration, the coefficient of d is crudely bounded between 0 and 1. However, the coefficient of ε is unbounded. This is the source of the error we see in the second test interval. This error can be removed by 'purifying' the argument, i.e., by perturbing the test argument x slightly to x' so that both x' and x'/3 are exact machine numbers and $\varepsilon = 0$. The following Fortran statements accomplish this on most computers.

```
Y = X/3.0E0
Y = (Y+X)-X
X = 3.0E0*Y.
```

The exceptions are those machines in which the active arithmetic registers are wider than the storage registers. On those machines storage of results must be forced at the completion of each arithmetic operation in the purification process. When purified arguments are used, the measured error is approximately

$$E = D - cd ,$$

where $0 \le c \le 1$.

A similar test procedure is derivable from the triple angle formula for the cosine function:

$$\text{cosine}(x) = \text{cosine}(x/3)\ [4\ \text{cosine}^2(x/3) - 3] .$$

Analysis similar to that for the sine function shows that x should be drawn from the interval $[(3m+1)\pi,(3m+3/2)\pi]$. The tests we recommend use $m = 2$. If argument purification is used, the resulting error expression is again

$$E = D - cd,$$

where E, D and d are defined analogously to the previous case, and where $0 \le c \le 2$.

Table IV contains results obtained using these tests, as implemented in the ELEFUNT package, on several different programs on several different computer systems. Tests labeled 3 are the cosine tests. The tests on the TI 59 were specially programmed and use only 50 arguments. We tested both the built-in sine function and one based on the algorithm described in section 2. We did not test the cosine function on the TI 59. Comparison of the IBM results with the master values shows how sharp these tests are. Large errors reported for other machines are probably caused by poor argument reduction or improper meshing of the sine and cosine computations.

Identity tests of the kind just described are seldom as discriminating as direct comparisons with higher precision calculations, but they can be surprisingly sharp and provide useful diagnostic information when properly designed and implemented. The test programs in ELEFUNT supplement accuracy tests based on identities with various other tests probing to see whether the function routines preserve important analytic properties of the function, whether the routines are robust, and how error conditions are handled. Thus, ELEFUNT exemplifies the physical examination approach to software testing that we advocate.

References.

[1] M. Abramowitz and I. A. Stegun, Handbook of Mathematical Functions with Formulas, Graphs, and Mathematical Tables, Nat. Bur. Standards Appl. Math. Series, 55, Washington D. C., 1964.

[2] C. W. Clenshaw, Mathematical Tables, Vol. 5, Chebyshev Series for Mathematical Functions, Her Majesty's Stationery Office, London, 1962.

[3] W. J. Cody, "Performance testing of function subroutines," AFIPS Conf. Proc., Vol. 34, 1969 SJCC, AFIPS Press, Montvale, N.J., 1969, pp. 759-763.

[4] ------, "The FUNPACK package of special function subroutines," TOMS 1, (1975), pp. 13-25.

[5] ------, "An overview of software development for special functions," Lecture Notes in Mathematics, 506, Numerical Analysis Dundee 1975, G. A. Watson (ed.), Springer Verlag, Berlin, 1976, pp. 38-48.

[6] ------, "Basic concepts for computational software," these proceedings

[7] ------ and W. Waite, Software Manual for the Elementary Functions, Prentice Hall, Englewood Cliffs, N.J., 1980.

[8] W. Gautschi, "Computational methods in special functions - a survey," Theory and Application of Special Functions, R. A. Askey (ed.), Academic Press, New York, 1975, pp. 1-98.

[9] C. Hammer, "Statistical validation of mathematical computer routines," AFIPS Conf. Proc., Vol. 30, 1967 SJCC, Thompson Book Co., Washington, D.C., 1967, pp. 331-333.

[10] J. F. Hart, E. W. Cheney, C. L. Lawson, H. J. Maehly, C. K. Mesztenyi, J.

R. Rice, H. C. Thacher, Jr. and C. Witzgall, Computer Approximations, Wiley, New York, 1968.

[11] C. Mesztenyi and C. Witzgall, "Stable evaluation of polynomials," NBS Jour. of Res. B, 71B (1967), pp. 11-17.

[12] A. C. R. Newbery and A. P. Leigh, "Consistency tests for elementary functions," AFIPS Conf. Proc., Vol. 39, 1971 FJCC, AFIPS Press, Montvale, N.J., 1971, pp. 419-422.

[13] J. R. Rice, "On the conditioning of polynomial and rational forms," Num. Math., 7 (1965), pp. 426-435.

[14] S. Walther, "A unified algorithm for the elementary functions," AFIPS Conf. Proc., Vol. 38, 1971 SJCC, AFIPS Press, Montvale, N.J., 1971, pp. 379-385.

PROGRAM CORRECTNESS AND MACHINE ARITHMETIC

by

T.J. Dekker

Abstract

The purpose of this paper is to give some insight in the construction of correct programs, especially numerical software, and in proving their correctness. After a brief survey of general program correctness axioms, the paper deals with various sets of axioms for machine arithmetic and some of its desirable features including a proposed standard for binary floating-point arithmetic. Moreover, a brief discussion is devoted to interval arithmetic. Finally, as an example of proving correctness, some algorithms for finding a zero of a real function in a real interval are considered.

Keywords: program correctness, machine arithmetic, floating-point system, interval arithmetic, zero finding, bisection, numerical software.

AMS - MOS Classification: 65G05, 65G10, 65H05, 68S24.

CR Classification 5.11, 5.15, 5.24.

Paper presented at CNR Seminar on Problems and methodologies in mathematical software production, Sorrento, Italy, November 1980.

Authors address: University of Amsterdam, Dept. of Mathematics,
Roetersstraat 15, Amsterdam, The Netherlands.

1. INTRODUCTION

Correctness of a program cannot in practice be proved by only testing the program. Indeed, to establish correctness in this way, one would have to try all possible input data, which would practically always require a prohibitive amount of time. This is well-known and mentioned by various authors, for instance, Dijkstra (1972), p.4, who considers as a simple example machine implementation of the multiplication of two integers.

In order to prove correctness of a program, one needs insight in its structure and in the properties of its constituent parts. This insight may be obtained by formulating suitable axioms which define the meaning of program structuring mechanisms and of constituent parts of programs such as elementary operations and basic statements.

On the other hand, correctness of a program cannot be proved without performing any testing of the program. Indeed, at least some tests are necessary to make sure that a program is a correct implementation of the intended algorithm (or set of algorithms), and that its text does not contain silly errors.

An adequate proof of correctness of a program must, therefore, not only contain a deduction of the correctness of the algorithms described in the program, but also an experimental verification based on the results of tests obtained for well-chosen sets of input data.

The deduction has to start from a suitable set of axioms and to be applied to the structuring mechanisms and the constituent parts of the program. It can only be successful, if the program has a clear structure, obtained by a systematic development of the program.

A well-known technique is the <u>top-down design</u> or <u>stepwise refinement</u>, i.e. a given problem is decomposed into precisely specified subproblems for which subsequently suitable (sub-)programs are developed. A proof of correctness consists of proving correctness of the decomposition into subproblems and of the corresponding subprograms. An experimental verification consists of tests of the subprograms separately as well as a test of the total program containing the subprograms.

A set of axioms is needed firstly to develop well-structured and correct programs and secondly to prove their correctness. In this paper we pay attention especially to the correctness of numerical software, and accordingly, to axioms for machine arithmetic.

In section 2, we mention some general program correctness axioms; in section 3, we deal with various sets of axioms for machine arithmetic; in section 4, we discuss some desirable features of machine arithmetic presented in literature, in section 5, interval arithmetic, and in section 6, as an example, algorithms for finding a zero of a real function in a real interval.

2. GENERAL PROGRAM CORRECTNESS AXIOMS

In this section, we deal with some general program correctness axioms, characterizing the effect of certain program statements. These axioms are to be used for designing well-structured and correct programs as well as for proving their correctness. We give only a concise set of axioms for some important statements, in order to give an idea of the theory.
For more details, see Hoare (1969), Dijkstra (1976), Alagić & Arbib (1978), Back (1979), Gries (1980), and the thorough theoretical treatment by De Bakker (1980). We use the notation of Alagić & Arbib (1978), which differs only slightly from that of Hoare (1969).

A program is written to solve a certain class of problems, defined by a class of permissible input data satisfying a certain initial condition and a class of output data - the required results - satisfying a certain final condition.

Let p denote the initial condition and q the final condition for a certain class of problems, and let S denote a statement or a sequence of statements or a program. We define two notions of correctness as follows.

2.1. Definition. S is partially correct with respect to initial condition p and final condition q, notation

$$\{p\}\ S\ \{q\}\ ,$$

if each execution of S which starts with input data satifying p and terminates after finitely many steps, yields output data satisfying q.

2.2. Definition S is totally correct with respect to initial condition p and final condition q, if each execution of S starting with input data satisfying p, terminates after finitely many steps and yields output data satisfying q.

We are now ready to present some program correctness axioms. Let S and T denote statements and p, q, r, b denote conditions.

2.3. Composition axiom

Let

$$S \; ; \; T$$

denote the composition of S and T, having the effect that first S and then T is executed. It is characterized by the following axiom:

if $\{p\}\ S\ \{q\}$ and $\{q\}\ T\ \{r\}$, then $\{p\}\ S\ ;\ T\ \{r\}$.

This axiom can be applied repeatedly to characterize the effect of a composition of more than two statements.

2.4. Assignment axiom

The simplest statement changing the value of one or more program variables, is the assignment statement. We consider an assignment statement of the form

$$v := f(v) \; ,$$

where v is a program variable (or an array of program variables in a language allowing assignment to an array), and f a function or expression which may or may not depend on v and on other program variables. We assume that evaluation of f causes no side effects on the program variables, and that v is a simple non-subscripted variable.

Moreover, we consider a final condition $q = q(v)$ depending on v and possibly also on other program variables. The effect of the assignment statement is characterized by the axiom:

$$\{q(f(v))\}\ v := f(v)\{q(v)\} \; .$$

Note that in $f(v)$ the old value of v is used, and in $q(v)$ the new value of v .

If we explicitly express that f and q may depend on other program variables which remain unchanged, we obtain the following formulation.
Let u denote a vector of program variables other than v ; consider an assignment statement of the form

$$v := f(u,v) ,$$

and a final condition $q = q(u,v)$; then the assignment statement satisfies

$$\{q(u, f(u,v))\} v := f(u,v) \{q(u,v)\} ,$$

provided that evaluation of f causes no side effects on the program variables.

2.5. Selection axiom

Consider the conditional statement

<u>if</u> b <u>then</u> S <u>else</u> T <u>fi</u> ,

where b is a Boolean expression whose evaluation does not cause side effects on the program variables. The effect of this statement is characterized by the following axiom:

if {p <u>and</u> b} S {q} and {p <u>and</u> <u>not</u> b} T {q} , then
{p} <u>if</u> b <u>then</u> S <u>else</u> T <u>fi</u> {q} .

A similar axiom has been formulated for the case statement, which is a statement selecting from several alternatives, see, for instance, Alagić & Arbib (1978).

2.6. Iteration axiom

The axioms mentioned above are equally valid for partial and total correctness; in other words, the statements considered terminate if all its constituent parts terminate. This does not hold for axioms on iteration statements.

We consider the iteration statement of the form

while b do S od ,

where again b is a Boolean expression without side effects. For this statement the following axiom holds:

if {p and b} S {p} then
{p} while b do S od {p and not b} .

Condition p is called the invariant assertion of the iteration, b the continuation condition and not b the stop criterion.
The iteration proceeds in the direction of the stop criterion not b leaving p invariant.

Invariant assertions play a fundamental role in the design of structured programs and in their correctness proofs. It is important to consider an appropriate invariant assertion for each iteration.

This iteration axiom does not say anything about termination (total correctness) of an iteration. Termination is proved by induction as follows: one defines a suitable (non-negative) integer function of the program variables whose value is an upper bound of the number of iteration to be performed; subsequently, one proves that the value of this function decreases in each iteration step and that the iteration terminates when the value of the function becomes zero (or negative). For details see, for instance, Alagić & Arbib (1978) or Gries (1980).

Similar axioms have been formulated for the repeat-until-statement (iteration with one or more steps) and the for-statement (iteration with a predetermined number of steps), and also for recursive procedures, see Hoare (1969) and De Bakker (1980).

2.7. Jump axiom

A statement containing one or more jumps (goto-statements) may have more than one exit. This makes it necessary to slightly expand the notation given above. Alagić & Arbib (1978) use the following notation for a statement S containing a jump to a label outside S

$\{p\}\ S\ \{q\}\ \{\ell : r\}$

which means that q is the final condition after normal termination of S and r the condition after termination of S by a jump to label ℓ. The goto-statement

goto ℓ

is characterized by the axiom

$\{p\}$ goto $\ell\ \{false\}\{\ell : p\}$,

and for the statement sequence

$S\ ;\ \ell : T$

the following axiom holds:

if $\{p\}\ S\ \{r\}\{\ell : r\}$ and $\{r\}\ T\ \{q\}\{\ell : r\}$ then
$\{p\}\ S\ ;\ \ell : T\ \{q\}$.

This axiom elucidates two kinds of jumps, namely the forward jump (goto-statement occurring in S) and the backward jump (goto-statement occurring in T). Only the second one causes an iteration. Here, condition r occurring at the label is an invariant assertion which plays the same role as the invariant assertion in the iteration axiom. In the design of a program containing jumps and for proving the correctness of such a program, it is very important to consider an appropriate invariant assertion at each label. A related set of axioms for multi-exit statements has been presented by Back (1979).

As is mentioned by several authors, one should avoid unnecessary use of goto-statements, because it makes programs hard to understand and hard to prove correct. Especially, one should not use the backward jump in a language containing well-structured iteration statements. A forward jump may, however, be appropriate as an exit from an iteration or a recursion.

3. AXIOMS FOR MACHINE ARITHMETIC

In this section, we deal with various sets of axioms for machine arithmetic. Machine arithmetic is (practically) always based on a floating-point representation of real numbers (the only reasonably adequate alternative seems to be a logarithmic representation).

A system of floating-point numbers can be described by a set of parameters, for instance as presented by Ford (1978) and by Brown & Feldman (1980), and by some set of axioms (depending on parameters), as presented, among many others, by Wilkinson (1963), Dekker (1971 & 1979), Brown (1980) and Holm (1980).

We firstly define floating-point number systems and parameters, sub equently describe four different sets of axioms which (apart from minor modifications) are taken from Dekker (1979) and are more or less common in literature, and conclude this section with a brief discussion of the parameter and axiom sets of Brown (1980) and Holm (1980).

3.1. Definition of floating-point systems

A floating-point number system is determined by four fundamental parameters, namely the base b , the mantissa length p , the minimal exponent emin and the maximal exponent emax . These parameters are integers satisfying $b > 1$, $p > 1$ and (for simplicity) $emin < -p$ and $emax > p$. We define a floating-point system in two variants, namely with or without denormalized numbers, as follows.

Definition. The floating-point system

$$\mathbb{F} = \mathbb{F}(b, p, emin, emax)$$

is the set of real numbers of the form

$$x = mx \times b^{ex - p} ,$$

where ex and mx are integers satisfying

$$emin \leq ex \leq emax , \quad |mx| < b^p .$$

A number x of this form is called a floating-point number, and the numbers

ex and mx are called exponent and mantissa (or fraction) of x respectively.

A mantissa is called normalized, if its magnitude is at least b^{p-1}.
A denormalized (or tiny) number is a non-zero floating-point number x having a non-normalized mantissa (i.e. $0 < |mx| < b^{p-1}$) and an exponent equal to the minimal possible value (i.e. $ex = emin$).

A floating-point system $\mathbb{F}$ as defined above is a system with denormalized numbers (namely, all elements x of $\mathbb{F}$ satisfying $0 < |x| < b^{emin-1}$ are denormalized). The smallest positive element of this system equals b^{emin-p}.

In many machine implementations, an additional requirement holds, stating that each non-zero mantissa must be normalized. We then obtain a floating-point system $\mathbb{F}$ without denormalized numbers. Its smallest positive element equals b^{emin-1}.

For a floating-point system defined in either way (with or without denormalized numbers), we define the following derived parameters, cf. Reinsch (1979):
the largest element

$$\lambda = (1-b^{-p})b^{emax} ;$$

the smallest positive element with normalized mantissa

$$\sigma = b^{emin-1} ;$$

the resolution

$$\Delta = b^{1-p} .$$

For each element x in $\mathbb{F}$ unequal to λ and $-\lambda$, let x^- denote its immediate predecessor and x^+ its immediate sucessor in $\mathbb{F}$; if, moreover $x \neq 0$, then its relative spacing $\rho(x)$ is defined by

$$\rho(x) = \frac{\max(x^+ - x \ , \ x - x^-)}{|x|} .$$

This quantity has lower bound b^{-p} and, for $|x| > \sigma$, its maximal value is equal to the resolution Δ . This property shows the importance of the parameter Δ .

A nice property of a floating-point system, with or without denormalized numbers, as defined above, is the symmetry, i.e. $0 \in \mathbf{F}$ and, if $x \in \mathbf{F}$, then also $-x \in \mathbf{F}$.

Machine implementations

A machine system of floating-point numbers, i.e. an actual implementation of a floating-point number system for a certain machine, may deviate from a floating-point system, with or without denormalized numbers, as defined above. We summarize the main possible deviations.

1. a machine system need not be symmetric; in particular, in a "base-b-complement" representation, a machine system may, for instance, contain $-b^{emax}$ but not its opposite value b^{emax} ;

2. a machine system may contain one or more "infinite" values and/or "indefinite" values to indicate, for instance, zero divide or unassigned value;

3. in a machine system, the set of stored values (which are stored in a memory location, for instance, by an assignment statement) may be a proper subset of the set of all possible values, obtained as result of arithmetical operations; thus, assignment may cause rounding or chopping to a stored value.

In order to cater for these deviations, and for possible anomalous arithmetic behaviour (in particular on underflow), Brown (1980) introduces a system of model numbers as a suitable (proper or improper) subset of a machine system (see section 3.3 below).

3.2. Arithmetic axioms

We consider the arithmetical operations addition, substraction, multiplication and division (denoted by $+, -, \times, /$) , the sign reversal (denoted by $-$) , and the arithmetic comparisons (denoted by $<, \leq, =, \neq, \geq, >$) . We disregard division by 0 .

For the corresponding implemented machine operations, we distinguish between normal operation, i.e. when the exponent limits do not influence the operation, and two kinds of abnormal operation, namely underflow and overflow, i.e. when the operation is influenced by the minimal exponent emin or the maximal exponent emax , respectively.

We assume that underflow and overflow either cause a trap (and deliver some machine dependent result which we leave unspecified), or set a flag and deliver some result as specified below in the axioms. In practice, the user can specify options if traps on underflow, overflow and/or zero-divide are enabled or not. In most situations it is recommended to have traps on overflow and zero-divide enabled (because mostly these events are desastrous) and a trap on underflow not enabled (because mostly on underflow a value near zero is delivered, and then underflow is often harmless; see also our remark below on gradual underflow).

We state four different sets of axioms of increasing strength which are more or less common in literature and fulfilled in several machine implementations. In the sequel, x and y denote elements of $\mathbb{F}$; moreover, $* \in \{+, -, \times, /\}$, i.e. $*$ denotes any one of the operations: addition, subtraction multiplication or division, and $\circledast$ denotes the corresponding machine operation.

Axiom set A (weak arithmetic)

This axiom set depends on the following parameters: the arithmetic precision or machine precision ε satisfying $0 < \varepsilon < 1$, the underflow threshold υ satisfying $\sigma \leq \upsilon < \varepsilon$, and the overflow threshold ω satisfying $\varepsilon^{-1} < \omega \leq \lambda$.

The machine operations corresponding to sign reversal and the arithmetic comparisons are exact and never lead to overflow or underflow; the machine operations corresponding to addition, subtraction, multiplication and division satisfy the following axioms.

A1) axiom on normal operation

On normal operation, the computed results of multiplication and division have a small relative error bounded by ε , whereas the computed results of addition and subtraction have a small absolute error bounded by ε times the sum of the magnitudes of the operands; in formulas (for division assuming $y \neq 0$, of course):

$$(3.2.1)\quad \begin{cases} |(x\hat{+}y) - (x+y)| \le (|x| + |y|)\varepsilon\ , \\ |(x\hat{-}y) - (x-y)| \le (|x| + |y|)\varepsilon\ , \\ |(x\hat{\times}y) - (x\times y)| \le |x \times y|\ \varepsilon\ , \\ |(x\hat{/}y) - (x/y)| \le |x\ /\ y|\ \varepsilon\ , \end{cases}$$

A2) axiom on underflow

Underflow does not occur when the exact result $x * y$ of an operation considered equals 0 or has a magnitude at least as large as υ .
A result on underflow not causing a trap either equals 0 ("flush to zero") or has the same sign as the exact result $x * y$ and magnitude at most as large as υ .

A3) axiom on overflow

Overflow does not occur when the exact result $x * y$ of an operation considered has a magnitude at most as large as ω . A result on overflow not causing a trap either has a finite magnitude at least as large as ω and the same sign as the exact result $x * y$, or some infinite value (with appropriate sign if the machine system contains distinct values $+\infty$ and $-\infty$).

Axiom A1 has been mentioned by Wilkinson (1963) for round-off with single precision accumulator (without guard digits). This axiom holds in every machine implementation of floating-point arithmetic for suitable ε of the same order of magnitude as the resolution Δ ; usually, one can take $\varepsilon = \Delta$, except possibly for division.

Axioms A2 and A3 hold in many machine implementations for suitable υ and ω of the same order of magnitude as σ and λ , respectively; usually, one can take $\upsilon = \sigma$ and $\omega = \lambda$.

Machine implementations

Machine implementations of floating-point arithmetic do not always satisfy axiom set A . The main deviations are the following.

1. in some implementations, arithmetic comparisons are evaluated by making the corresponding comparison of $x \hat{-} y$ and 0 ; consequently, arithmetic comparisons may yield incorrect results and even lead to underflow or overflow;

2. sign reversal may lead to overflow or underflow, if the machine system of floating-point numbers is not symmetric;

3. in some implementations (for CD Cyber machines), <u>operand underflow</u> may occur, i.e. a non-zero operand of multiplication or division with magnitude smaller than $\upsilon(=\sigma)$ is replaced by 0, so that the operation yields a computed result 0 or leads to zero-divide; operand underflow is an anomalous kind of underflow, which does not satisfy axiom A2 .

<u>Axiom set B (strong arithmetic)</u>

This axiom set is obtained from axiom set A by replacing axiom A1 by the following.

B1) <u>axiom on normal operation</u>

On normal operation, the computed results of addition and subtraction as well as multiplication and division have a small relative error bounded by ε ; in formula:

$$|(x \hat{*} y) - (x * y)| \leq |x * y| \varepsilon \; . \tag{3.2.2}$$

Axiom B1 has been mentioned by Wilkinson (1963) for round-off with double precision accumulator. In fact, only one guard digit is needed to achieve strong arithmetic (see also remark on machine implementation at axiom set D below). Clearly, axiom set B implies axiom set A with the same values for ε, υ and ω .

<u>Axiom set C (faithful arithmetic)</u>

This axiom set is obtained from axiom set A by substituting $\upsilon = \sigma$ and $\omega = \lambda$, and replacing axioms A1 and A2 by the following.

C1,2) <u>axiom on normal operation and underflow</u>

On normal operation and on underflow not causing a trap, the computed result of a machine operation corresponding to addition, subtraction, multiplication or division is <u>faithful</u>, i.e. it equals either the largest element of $\mathbb{F}$ not larger, or the smallest element of $\mathbb{F}$ not smaller than the exact result; in other words:

> if $x * y \in \mathbb{F}$, then $x \hat{*} y = x * y$;
> if $a < x * y < a^{+}$, where a and a^{+} are successive elements of $\mathbb{F}$, then $x \hat{*} y$ equals either a or a^{+} .

Moreover, underflow does not occur when the exact result $x * y$ equals 0 or has a magnitude at least as large as $\upsilon = \sigma$.

Clearly, axiom set C implies axiom set B with $\varepsilon = \Delta$, $\upsilon = \sigma$ and $\omega = \lambda$. In axiom set C , a result on overflow not causing a trap equals either $\pm \lambda$ with appropriate sign or some infinite value; this follows from the fact that axiom A3 holds with $\omega = \lambda$.

Gradual underflow

In faithful arithmetic, the results on underflow (not causing a trap) are substantially different in systems with or without denormalized numbers. In a system without denormalized numbers, these results are either 0 (flush to 0), or $\pm \sigma$ with appropriate sign; in either case, the absolute error is bounded by σ ; in a system with denormalized numbers, these results have an absolute error bounded by $\sigma\Delta = b^{emin-p}$ for multiplication and division, and are even exact for addition and subtraction. Consequently, the results of addition and subtraction have a small relative error satisfying formula (3.2.2) with $\varepsilon = \Delta$, not only on normal operation, but also on underflow (not causing a trap). This nice behaviour on underflow in a system with denormalized numbers is called gradual underflow, cf. Kahan & Palmer (1979), and Coonen (1981). It has been included in the proposed standard for binary floating-point arithmetic, see Stevenson a.o. (1981).

Axiom set D (proper rounding arithmetic)

This axiom set is obtained from axiom set C by adding the following requirements in axiom C1,2:

on normal operation, the computed result of any of the operations considered is properly rounded, i.e. it equals an element of $\mathbb{F}$ nearest to the exact result;

the same holds on underflow not causing a trap, except possibly when the exact result is between two successive elements a and a^+ of $\mathbb{F}$ and either a or a^+ equals 0 ; in this case $x \,\hat{*}\, y$ may be equal to either a or a^+ .

The exception on underflow has the purpose to allow, for instance, flush to 0 on underflow.

Clearly, D implies C ; also D implies B with $\varepsilon = \frac{1}{2}\Delta$, $\upsilon = \sigma$, $\omega = \lambda$.

Machine implementation

Arithmetic satisfying any of these axiom sets is easy to implement. In particular, to achieve proper rounding arithmetic, only two guard digits are needed (cf. Knuth (1969), section 4.2.1 exercise 5).

The most ideal arithmetic is unbiased proper rounding arithmetic. This satisfies axiom set D with an extra "tie-breaking rule" to be applied when $x * y$ lies halfway between two successive elements a and a^+ of $\mathbb{R}$; usually, the tie-breaking rule chosen is round to even, i.e. $x \hat{*} y$ equals that element a or a^+ which has an even p - digit mantissa. To achieve this, one needs not only two guard digits, but also a "sticky" bit indicating if beyond the guard digits some non-zero digits were discarded, cf. Stevenson a.o. (1981).

3.3. Brown's model and axiom set

Brown (1980) presents a model of floating-point computation which depends on fundamental parameters as mentioned above and includes a set of axioms. An important feature is the distinction between machine numbers and model numbers. A system of model numbers is defined as a system of floating-point numbers without denormalized numbers, as in section 3.1 above. This system is a (proper or inproper) subset of a system of machine numbers.

Definitions.

Let $\mathbb{F}$ denote the system of model numbers and let λ, σ, Δ denote its derived parameters as defined in section 3.1 above.

A model interval is a closed interval $[x,y]$, where x and y are model numbers satisfying $x \leq y$. If r is any real number r satisfying $|r| \leq \lambda$, then r' denotes the smallest model interval containing r ; i.e. if r is a model number, then $r' = [r,r] = \{r\}$, otherwise $r' = [x,x^+]$, where x and x^+ are two consecutive model numbers satisfying $x < r < x^+$.

Similarly, for any closed real interval I contained in $[-\lambda,\lambda]$, I' denotes the smallest model interval containing I (in particular if $I = [r,s]$ and both r and s are model numbers, then $I' = I$).

Let I and J be two model intervals. Arithmetical operations on these intervals are defined as follows:

$$I * J = \{r * s \mid r \in I \text{ and } s \in J\} ,$$

where for division it is assumed that J does not contain 0 ; moreover,

$$-I = \{-r \mid r \in I\} .$$

Some of the main axioms for Brown's model are the following.

Axiom 1

Let x and y be machine numbers satisfying $|x| \leq \lambda$ and $|y| \leq \lambda$; let the interval $x' * y'$ be contained in $[-\lambda,\lambda]$ (i.e. normal operation and underflow are considered, but not overflow). Then, except possibly for division:

$$x \mathbin{\hat{*}} y \in (x' * y')' .$$

The same condition or a slightly weaker condition holds for division, assuming (of course) that 0 is not an element of y' .

Axiom 2 (special case)

Let x be a machine number satisfying $|x| \leq \lambda$; then

$$\hat{-}\, x \in -(x') .$$

In particular, sign reversal is exact when x is a model number.

Consequences

Brown's axiom set is valid for any machine system provided that the parameters defining the system of model numbers are suitably chosen. Especially thanks to the distinction between machine numbers and model numbers, the set of axioms can cater for various known anomalies in machine systems and machine arithmetic (see our remarks on machine implementations in section 3.1 and at axiom set A).

An important special case is obtained when the set of model numbers is equal to the set of machine numbers. In this case, Brown's axiom set implies our axiom set C (faithful arithmetic), apart from possible results on overflow.

3.4. Holm's axiom set

Holm (1980) presents a set of 15 axioms with the following main features.

The results of arithmetical operations are obtained from the corresponding exact results by applying two cropping (i.e. chopping or rounding) functions. These cropping functions are monotonic (i.e. non-decreasing for increasing argument), anti-symmetric (i.e. if $c = c(x)$ is a cropping function, then $c(-x) = -c(x)$), and faithful.

For multiplication and division only one cropping function is needed; for addition and subtraction two cropping functions are used, depending on whether the operands have the same sign or opposite signs. Given these two cropping functions, all results are uniquely determined. Various cropping functions can be defined for various number of guard digits (≥ 1) in the accumulator.

Consequences
The arithmetic is faithful (satisfies our axiom set C) and deterministic (results are uniquely determined by two given cropping functions). Moreover, addition and multiplication are commutative.

4. DESIRABLE FEATURES OF FLOATING-POINT ARITHMETIC.

Desirable features of floating-point systems and arithmetic have been formulated by various authors, and have been and are still being discussed by the IFIP working group 2.5 on numerical software, see Dekker (1977), Reinsch (1979) and Hull (1979). Moreover, the floating-point working group of the Microprocessor standards subcommittee of the IEEE Computer Society has proposed a standard for binary floating-point arithmetic, described by Stevenson a.o. (1981), see also Kahan & Palmer (1979), Coonen (1980 & 1981), Cody (1981) and Hough (1981).
We briefly describe some of the desirable features mentioned and the proposed standard.

Reinsch' principles and preferences
Some of the main principles and preferences mentioned by Reinsch (1979) are the following.

Preference 1.1. $\mathbb{F}$ is to be symmetric, i.e. 0 is in $\mathbb{F}$ and if x is in $\mathbb{F}$ then also $-x$.

Preference 1.2. $\mathbb{F}$ is to be balanced , i.e. $\sigma \times \lambda$ should have order of magnitude one.

Principle 2.1. The hardware reference manual must provide a precise and complete algorithmic description of the implemented floating-point operations.

Preference 2.2. The rounding error in the floating-point operations must be locally unbiased or, at least, have a negligeable bias compared with the "root-mean-square" error.

Principle 2.3. and preference 2.4. On normal operation, the floating-point operations must satisfy axiom set A or, preferably, axiom set B described above. The hardware reference manual must mention the value of the corresponding arithmetic precision ε .

Principles 2.5 and 2.6. Floating-point operations must preserve monotonicity and anti-symmetry of the corresponding mathematical operations.

Principles 3.1 and 3.2. Each occurrence of overflow or underflow must be reported. More generally, a floating-point operation must never deliver an incorrect or undefined result without an error indication.

Principle 4.1. The arithmetic comparisons must always yield the correct result and never yield underflow or overflow.
For details see Reinsch (1979).

Hull's desirable floating-point arithmetic
Hull (1979) formulates some desirable features for an ideal arithmetic system and for elementary functions.
An ideal arithmetic system should be complete, simple and flexible.

Complete means that the programmer knows what will happen under any circumstance. Simplicity leads to the conclusions, for example, that the base should be 10 and the number representation should be sign-magnitude. Both completeness and simplicity lead to the conclusion that on normal operation arithmetic should be properly rounding (axiom set D), with an additional tie-breaking rule, such as: round-to-even, mentioned above.

Flexibility leads to the conclusion that the programmer should be allowed to change the precision and range of program variables and also to specify dynamically (i.e. at run time) in which precision and range arithmetic operations are to be performed.

Moreover, some desirable features on detection and handling of exceptions (underflow, overflow, zero-divide, etc.), and on the evaluation of elementary functions are mentioned. For details, see Hull (1979).

Standard for binary floating-point arithmetic

Some main features of the standard proposed by Stevenson a.o. (1981), and adopted by the working group of the IEEE Computer Society mentioned above, are as follows.

The base b of the floating-point system equals 2 .

An implementation conforming to the proposed standard contains one or more distinct floating-point systems with different precision and range, according to some defined formats. These distinct systems are, in our notation, as follows:

a single-precision system

$$\mathbb{F}_1 = \mathbb{F}(2, 24, -125, 128) ,$$

a double-precision system

$$\mathbb{F}_2 = \mathbb{F}(2, 53, -1021, 1024) ,$$

and, moreover, a single-extended and a double-extended precision system which have a larger precision and range than $\mathbb{F}_1$ or $\mathbb{F}_2$, respectively.

Each of these systems is a system with denormalized numbers. The systems also contain bit configurations for infinity and for signaling arithmetic exceptions (called "NaN" , i.e. "not a number").

The elements of $\mathbb{F}_1$ and $\mathbb{F}_2$ with a normalized mantissa are represented with implicit first bit, i.e. the first bit, which always equals 1 , is not stored in the memory location representing the number. Two special (out of range) exponent bit configurations are used to represent 0 , denormalized numbers, infinity and NaN's . Consequently, 32 bits are needed for the elements of $\mathbb{F}_1$ (1 sign bit, 8 exponent bits and 23 explicit mantissa bits), and 64 bits for the elements of $\mathbb{F}_2$ (1 sign bit, 11 exponent bits and 52 explicit mantissa bits).

The minimal configuration conforming to the standard contains $\mathbb{F}_1$; recommended larger configurations contain either $\mathbb{F}_1$ and a single-extended system or $\mathbb{F}_1$, $\mathbb{F}_2$ and a double-extended system.

The arithmetic conforming to the proposed standard satisfies our axiom set D (proper rounding) with round-to-even as tie-breaking rule. Besides proper rounding, three different kinds of directed rounding must be available as user-selectable options, namely truncation (round towards 0), rounding down (towards $-\infty$) and rounding up (towards $+\infty$) . These directed rounding arithmetics satisfy our axiom set C , and are important features for conversion to integers and for interval arithmetic.

In each kind of arithmetic (proper or directed roundings), there are two possible modes for handling infinity:

the projective mode (the default mode) handles $+\infty$ and $-\infty$ as equal values which are unordered with respect to the finite numbers;

the affine mode handles $+\infty$ and $-\infty$ as distinct values which are ordered such that $-\infty$ is smaller and $+\infty$ is larger than each finite number.

For details, see Stevenson a.o. (1981), Coonen (1980 & 1981), Cody (1981) and Hough (1981).

5. INTERVAL ARITHMETIC

We here present a brief survey of the main features of interval arithmetic. The purpose of interval arithmetic is to keep track of lower bound and upper bound of the results obtained in a numerical computation.

Moore intervals

The simplest form of interval arithmetic is presented by Moore (1966 & 1979).

A Moore interval $A = [\underline{a}, \bar{a}]$, where $\underline{a}$ and $\bar{a}$ are real numbers satisfying $\underline{a} \leq \bar{a}$, is by definition equal to the closed interval whose end points are $\underline{a}$ and $\bar{a}$. The end points $\underline{a}$ and $\bar{a}$ are called lower bound and upper bound of A , respectively.

The arithmetical operations on intervals are defined as follows:

for $* \in \{+, -, \times, /\}$
and intervals $A = [\underline{a}, \bar{a}]$ and $B = [\underline{b}, \bar{b}]$,

$$A * B = \{z \mid z = x * y \text{ and } \underline{a} \leq x \leq \bar{a} \text{ and } \underline{b} \leq y \leq \bar{b}\} ,$$

except that division is not defined when $\underline{b} \leq 0 \leq \bar{b}$.

Thus we obtain the following formulas expressed in lower and upper bounds:

$$[\underline{a}, \bar{a}] + [\underline{b}, \bar{b}] = [\underline{a} + \underline{b} , \bar{a} + \bar{b}] ;$$
$$[\underline{a}, \bar{a}] - [\underline{b}, \bar{b}] = [\underline{a} - \bar{b} , \bar{a} - \underline{b}] ;$$
$$[\underline{a}, \bar{a}] \times [\underline{b}, \bar{b}] = [\min(V), \max(V)],$$

where V is the set of four real values defined by

$$V = \{\underline{a}\,\underline{b} , \underline{a}\bar{b} , \bar{a}\underline{b} , \bar{a}\bar{b}\} ;$$

finally, when $\underline{b} > 0$ or $\bar{b} < 0$,

$$[\underline{a}, \bar{a}] / [\underline{b}, \bar{b}] = [\underline{a}, \bar{a}] \times [^1/\bar{b}, {}^1/\underline{b}] .$$

Moreover, the sign reversal is defined by

$$-[\underline{a}, \bar{a}] = [-\bar{a}, -\underline{a}] .$$

For intervals of length 0 (i.e. $\underline{a} = \bar{a}$) , the arithmetic coincides with ordinary real arithmetic, where each real number a corresponds to the interval $[a, a]$. Interval arithmetic as defined above satisfies the associative and commutative laws both for addition and multiplication, but not the distributive law. Instead, it satisfies the subdistributive law, i.e. for any intervals A, B, C , we have

$$A \times (B+C) \subset A \times B + A \times C .$$

For instance, if $A = [0, 1]$, $B = [1, 1]$, $C = [-1, -1]$, then

$$A \times (B+C) = [0, 1] \times [0, 0] = [0, 0] ,$$
$$A \times B + A \times C = [0, 1] + [-1, 0] = [-1, 1] ,$$

which shows that, for this example, $A \times (B+C)$ is a proper subset of $A \times B + A \times C$.

Moreover, the interval arithmetic yields intervals at each stage, whose end points are lower and upper bound of the corresponding exact results, provided that the initial data are intervals satisfying this same property. The bounds obtained may, however, be rather pessimistic, in particular, when a dependency occurs in a calculation, i.e. when the same variable occurs more than once.

The dependency width is the difference of the interval length obtained by standard interval arithmetic and the length of interval obtained when the dependencies would be taken into account, cf. Ris (1975).

For instance, if $A = [1, 2]$, then

$$A - A = [-1, 1] \ , \ A/A = [0.5, 2] \ ;$$

taking into account the dependency, however, the results should be $[0, 0]$ and $[1, 1]$ respectively; so, the dependency width in these cases equals 2 and 1.5 , respectively.

The resulting dependency width is the main shortcoming of standard interval arithmetic. There are many techniques to reduce the dependency width, see various publications on interval arithmetic such as Moore (1966 & 1979) and Nickel (1975).

A variant of the Moore intervals are the triplex intervals introduced by Nickel, see for instance Apostolatos a.o. (1968).
A triplex interval is a combination of a Moore interval and a main value which is a real value between the end points of the interval.

Kahan intervals

Besides Moore intervals as defined above, Kahan introduces intervals including infinity. The real line is extended to the projective real line which is homeomorphic (i.e. topologically equivalent) to a circle and contains the set of real numbers and one infinite value (in other words $-\infty = +\infty$, similarly as in the projective mode of the floating-point standard mentioned above).

Thus, the following six kinds of intervals are obtained, where a and b denote finite real values:

1) $A = [a, b] \ , \ a \leq b \ ;$

this is the Moore interval as defined above;

2) $A = [a, \infty]$;

this is the half-infinite interval containing the real numbers $\geq a$ and infinity;

3) $A = [\infty, b]$;

this is the half-infinite interval containing the real numbers $\leq b$ and infinity;

4) $A = [a, b]$, $a > b$;

this interval contains the real numbers $\geq a$, infinity, and the real numbers $\leq b$; in the projective line topology it contains infinity in its interior;

5) $A = [\infty, \infty]$;

this interval contains only infinity.

6) $A = \Omega$;

this contains all real numbers and infinity (in other words, Ω may be used to denote "indefinite").
For details see Kahan (1968) and Laveuve (1975).

Implementations of interval arithmetic

Implementations of interval arithmetic have to take into account the inexactness of floating-point computation. Consequently, the lower bounds have to be calculated by rounding down (towards $-\infty$) and the upper bounds by rounding up (towards $+\infty$). The main values in triplex arithmetic are calculated by the ordinary machine arithmetic (preferably proper rounding).

Packages for interval arithmetic (both for Moore intervals and for triplex intervals) have been produced by Yohe and others, see Yohe (1979).
Moreover, triplex interval arithmetic has been implemented in the form of

an extension to Algol 60, called triplex Algol 60, see Apostolatos a.o. (1968). The proposed floating-point standard mentioned above contains valuable features for implementing interval arithmetic, in particular, the directed roundings.

Applications of interval arithmetic

Interval arithmetic may be very useful for certain calculations. It yields rigorous error bounds at an extra cost which need not be more than 50 or 100 procent. The error bounds obtained may, however, be very pessimistic, mainly due to the dependency width. In view of this, special techniques are often needed, in order to take into account the dependencies occurring in the formulas used. Such special techniques exist, for instance, for the evaluation of polynomials and rational functions, for the solution of (systems of) linear or non-linear algebraic equations, for optimization methods and for the solution of ordinary differential equations, see Moore (1966 & 1979) and Nickel (1975) and other literature on the subject.

A general technique, applicable to systems of linear or non-linear algebraic equations, is the following.
Calculate an approximate solution of a given system using ordinary floating-point arithmetic; subsequently, use interval arithmetic to calculate the residual corresponding to this approximate solution and a rigorous error bound. If needed, one can use this information to calculate an improved approximate solution (iterative refinement), see for instance Wilkinson (1963).

6. EXAMPLE: FINDING A REAL ZERO.

We consider the problem of finding a zero of a continuous real-valued function defined on a real interval. We assume that the values of the function at the end points of the interval are neither both positive nor both negative (i.e. these values either have opposite sign or at least one of them equals 0). Hence, according to classical analysis, the function must have a zero in the interval.

The required result is a pair of real numbers in the given interval which include a zero of the function and have a difference not larger than a given tolerance.

We consider some iterative algorithms to solve this problem and focus attention on their performance in floating-point arithmetic satisfying any of the axiom sets A, B, C, D given above.

We assume that underflow does not cause a trap (underflow trap not enabled), and (for simplicity) that no overflow occurs in the calculations considered (this can be easily checked when overflow trap is enabled).

In our error analysis, we disregard effects of errors in computing values of the given function. Thus, we assume that the function is given (exactly) as a routine yielding a value in $\mathbb{F}$ for each argument value in the intersection of $\mathbb{F}$ and the given interval. The effect of errors in the given function can be handled by standard techniques.

Furthermore, we assume that the tolerance is given not as a constant but as a linear function of the form

$$t(x) = \max(\rho \,\hat{\times}\, |x|, \alpha) , \tag{6.1}$$

where ρ and α are certain positive constants of the same order of magnitude as the machine precision and the underflow threshold, respectively (lower bounds for these constants are specified below).

<u>Notations</u>

Let f denote the given function, and $x0, y0 \in \mathbb{F}$ be the end points of the given interval.
For $x, y \in \mathbb{F}$, let $J(x,y)$ denote the closed real interval $[x, y]$ if $x \le y$ or $[y, x]$ otherwise, and $\hat{J}(x,y)$ the intersection of $J(x,y)$ and $\mathbb{F}$. Moreover, let $\mathrm{sign}(x)$ be the function yielding the value $+1, 0, -1$ for positive, zero, negative real argument x, respectively.

<u>Formulation of the problem</u>

The given data are two real numbers $x0, y0 \in \mathbb{F}$, a function f yielding a value in $\mathbb{F}$ for each argument value in $\hat{J}(x0,y0)$ and a function t of the form (6.1). The given data $x0$, $y0$ and f must satisfy the <u>initial condition</u>

$$\mathrm{sign}(f(x0)) \times \mathrm{sign}(f(y0)) \le 0 . \tag{6.2}$$

<u>Remark.</u> We use this condition instead of the simpler, mathematically equivalent, condition $f(x0) \times f(y0) \le 0$, because in floating-point arithmetic

the corresponding condition $f(x0) \hat{\times} f(y0) \leq 0$ may yield a wrong result or lead to overflow or underflow; condition (6.2) performed in machine arithmetic amounts to an exact arithmetic comparison using only the sign of $f(x0)$ and $f(y0)$ *end of remark.*

The required results are $x, y \in \hat{J}(x0,y0)$ satisfying the final condition

$$(6.3) \qquad \begin{cases} \text{sign}(f(x)) \times \text{sign}(f(y)) \leq 0 , \\ |y \hat{-} x| \leq t(x) , \\ |f(x)| \leq |f(y)| . \end{cases}$$

In other words, x and y are the end points of a sufficiently small interval containing a zero of f . The latter condition is added for definiteness; accordingly, we call x the "best" approximation found, and y the "contrapoint", i.e. a value found during the iteration such that x and y include a zero of f and y is nearest to x .

Algorithms

To solve this problem, we consider iterative algorithms of the following type. In each iteration step a current interval is reduced to a proper subinterval leaving a certain assertion invariant.

Let b and c denote variables whose values are the end points of the current interval. Then the invariant assertion is

$$(6.4) \qquad \begin{cases} \text{sign}(f(b)) \times \text{sign}(f(c)) \leq 0 , \\ |f(b)| \leq |f(c)| , \end{cases}$$

and the stopping criterion is

$$(6.5) \qquad |c \hat{-} b| \leq t(b) .$$

(Here again the latter condition of (6.4) is added for definiteness; deletion of this condition may lead to a slightly simpler algorithm; we do not persue this further.).

In particular, we consider the bisection algorithm and a class of bounded interpolation algorithms, i.e. algorithms combining some interpolation formula with bisection.

Two algorithms of this class, using linear interpolation and a three-point rational interpolation formula, were presented by Bus & Dekker (1975), another related algorithm by Brent (1971).

We consider two numerically different (but mathematically equivalent) versions of bisection. In the first version, Z1 , the next iterate in each step is calculated by the formula

$$(6.6) \qquad m1 = (b \hat{+} c) \hat{/} 2 ,$$

in the second version, Z2 , by the formulas

$$(6.7) \qquad \begin{cases} w = c \hat{-} b , \\ h = w \hat{/} 2 , \\ m2 = b \hat{+} h . \end{cases}$$

Let Z3 denote any bounded interpolation algorithm.
Let, in each step of Z3 ,

$$(6.8) \qquad \begin{cases} r = \mathrm{sign}(w) \times t(b) \hat{/} 2 , \\ s = b \hat{+} r . \end{cases}$$

We assume that in each step of Z3 the next iterate either equals a value, i, according to a certain interpolation formula if this value is (approximately) in the interval J(s, m2) , or otherwise equals s or m2 . Different algorithms of the class considered may however use different interpolation formulas and different strategies to select a value i or s or m2 as next iterate. For details, see Brent (1971), Bus & Dekker (1975), or Dekker (1979).

We now are ready to formulate some theorems for algorithms Z1, Z2, Z3, which are the same as those presented in Dekker (1979), apart from minor modifications in the formulation of the axiom sets.

<u>Theorem 1</u>
Let algorithms Z1, Z2, Z3 be performed in arithmetic satisfying any of the sets of axioms A, B, C, D .
Let Δ be the resolution of $\mathbb{F}$, let ε be the arithmetic precision and υ the underflow threshold in axiom sets A and B , and let $\varepsilon = \Delta$ in

axiom set C , $\varepsilon=\frac{1}{2}\Delta$ in axiom set D and $\upsilon = \sigma$ in axiom sets C and D .

We assume that $0 < \varepsilon < \frac{1}{8}$ and that the tolerance function t has the form (6.1) , where the constants ρ and α satisfy the following conditions:

$$\alpha > 2\upsilon \;/\; (1-8\varepsilon) \;,$$
$$\rho > C_\varepsilon \;/\; (1-8\varepsilon) \;,$$

where C_ε is a constant specified by the following table for the various algorithms and axiom sets

C_ε	A	B	C	D
Z1	6ε	4ε	$3\varepsilon(=3\Delta)$	$4\varepsilon(=2\Delta)$
Z2, Z3	6ε	2ε	$2\varepsilon(=2\Delta)$	$2\varepsilon(=\Delta)$

Moreover, we assume that no overflow or division by 0 occurs, and that underflow does not cause a trap.

Within these assumptions, the theorem states for the algorithms mentioned, that each iteration step yields a next iterate between the end points, b and c , of the current interval.

Remark. We prove the theorem only for algorithm Z2 in axiom sets B, C and D . The corresponding results for algorithm Z1 and Z3 , and those for axiom set A , are derived in a similar way, where for algorithm Z3 it is essential that a value i of a certain interpolation formula is delivered as next iterate only when it is approximately within the interval $J(s, m2)$. The factor $1 - 8\varepsilon$ mentioned in the theorem, smaller than necessary for algorithms Z1 and Z2 , is chosen to fit the proof for algorithm Z3 . For details see Dekker (1979).

Proof for algorithm Z2 in axiom sets B, C and D.

We first consider axiom set B .

In view of the stopping criterion (6.5) , the iteration is continued only if

$$(6.9) \qquad |w| = |c \mathbin{\hat{-}} b| > t(b) = \max(\rho \mathbin{\hat{\times}} |b|, \alpha) \;;$$

hence, in particular

$$(6.10) \qquad |w| > \frac{C_\varepsilon |b|(1-\varepsilon)}{1-8\varepsilon} = |b| \frac{2\varepsilon(1-\varepsilon)}{1-8\varepsilon} .$$

Without loss of generality, we assume $w > 0$; otherwise, we reverse the sign of b and c and all other argument values, and consider the corresponding step for the function g defined by $g(x) = f(-x)$.
We now show that $b < m2 < c$.

On normal operation

On normal operation, it follows from formula (6.7) that m2 satisfies

$$\frac{w(1-\varepsilon)^2}{2} + b - |b|\varepsilon \le m2 \le \frac{w(1+\varepsilon)^2}{2} + b + |b|\varepsilon .$$

Hence, using (6.10) ,

$$m2 - b > |b|\{ \frac{\varepsilon(1-\varepsilon)^3}{1-8\varepsilon} - \varepsilon\} \ge 0 ;$$

moreover, since $w = c \hat{-} b$,

$$c - b \ge \frac{w}{1+\varepsilon} ,$$

so that, again using (6.10) ,

$$c - m2 = c - b + b - m2 \ge w\left[\frac{1}{1+\varepsilon} - \frac{(1+\varepsilon)^2}{2}\right] - |b|\varepsilon$$

$$> |b|\left\{\frac{2-(1+\varepsilon)^3}{(1+\varepsilon)} \times \frac{\varepsilon(1-\varepsilon)}{1-8\varepsilon} - \varepsilon\right\} \ge 0 .$$

This proves that $b < m2 < c$ on normal operation.

On underflow

On underflow the stopping criterion (6.5) or (6.9) yields, that during the iteration w satisfies

$$|w| = |c \hat{-} b| > \alpha > \frac{2\upsilon}{1-8\varepsilon} .$$

Hence, it follows from formula (6.7) , that both w and $h = w \hat{/} 2$, have a magnitude larger than υ , so that underflow cannot occur in the calculation of these quantities. So, underflow can only occur in the calculation of $f(x)$, where it is completely harmless, and in the calculation of $m2 = b \hat{+} h$.

Since on underflow the error is smaller than υ , we have

$$b + h - \upsilon < m2 < b + h + \upsilon .$$

Hence,

$$m2 - b > h - \upsilon \geq \frac{w(1-\varepsilon)}{2} - \upsilon \geq \frac{\upsilon(1-\varepsilon)}{1 - 8\varepsilon} - \upsilon > 0 ,$$

$$c - m2 = c - b + b - m2 > \frac{w}{1 + \varepsilon} - h - \upsilon$$

$$\geq w\left[\frac{1}{1+\varepsilon} - \frac{1+\varepsilon}{2}\right] - \upsilon \geq \frac{2 - (1+\varepsilon)^2}{1+\varepsilon} \times \frac{\upsilon}{1 - 8\varepsilon} - \upsilon > 0 ,$$

which proves that $b < m2 < c$ when underflow occurs.

This completes the proof for alogorithm Z2 in axiom set B .

The corresponding results in axiom sets C and D follow by observing that these axiom sets both imply axiom set B with $\varepsilon = \Delta$ or $\varepsilon = \frac{1}{2}\Delta$, respectively, and with $\upsilon = \sigma$. □

Theorem 2

The lower bounds for ρ and α in theorem 1 are best possible in axiom sets A and B within a factor $1 + O(\varepsilon)$.

Proof

We prove this theorem by means of the following examples which show that ρ must be larger than C_ε and α larger than 2υ .

Examples concerning the lower bound for ρ

Take $b = 1$, $c = 1 + 4\varepsilon$.

Within the limitations of axiom set A , we can have for algorithm Z1

$$b \ddagger c = 2 + 6\varepsilon , \quad m1 = 1 + 4\varepsilon = c , \quad w = 6\varepsilon ,$$

and for algorithms Z2 and Z3

$$w = 6\varepsilon , \quad h = 3\varepsilon , \quad m2 = 1 + 4\varepsilon = c ,$$

which shows that ρ must be larger than 6ε.

Similarly, with the same values for b and c, we can have in axiom set B for algorithm Z1

$$b \mathbin{\hat{+}} c = 2 + 6\varepsilon \ , \quad m1 = 1 + 4\varepsilon = c \ , \quad w = 4\varepsilon \ ,$$

showing that ρ must be larger than 4ε ; for algorithms Z2 and Z3 , taking $b = 1$, $c = 1 + 2\varepsilon$, we can have in axiom set B

$$w = 2\varepsilon \ , \quad h = \varepsilon \ , \quad m2 = 1 + 2\varepsilon = c \ ,$$

showing that ρ must be larger than 2ε .

Examples concerning the lower bound for α

Take $b = 0$, $c = 2\upsilon$.

Within the limitations of both axiom sets A and B (with flush to 0 on underflow), we can have for algorithm Z1

$$b \mathbin{\hat{+}} c = 2\upsilon(1-\varepsilon) \ , \quad m1 = 0 = b \ , \quad w = 2\upsilon \ ,$$

and for algorithms Z2 and Z3

$$w = 2\upsilon \ , \quad h = \upsilon(1-\varepsilon) \ , \quad m2 = 0 = b \ ,$$

showing that in all cases α must be larger than 2υ . □

Theorem 3

If the conditions of theorem 1 hold true, then algorithms Z1, Z2, Z3 terminate in a finite number of steps.

Proof. According to theorem 1 , each iteration step yields a next iterate between the end points of the current interval. From this the result immediately follows, because $\mathbb{F}$ is finite. □

Conjecture

Theorem 3 is not of much value, because it does not yield a useful upper bound for the number of iterations required.

The author conjectures that this number is of the same of order of magnitude as the corresponding number of iterations for the same algorithms performed in exact arithmetic, if the conditions of theorem 1 are satisfied and, moreover, ρ and α are larger than the bounds given in theorem 1 times a certain constant larger than 1 .

7. REFERENCES

Program correctness

S. Alagić & M.A. Arbib, *The design of well-structed and correct programs;* Springer-Verlag (1978).

R.J.R. Back, *Exception handling with multi-exit statements;* report IWI 125/79, Math. Centre, Amsterdam (1979).

J. de Bakker, *Mathematical theory of program correctness*; Prentice-Hall (1980).

E.W. Dijkstra, *Notes on structured programming* in O.J. Dahl, E.W. Dijkstra, C.A.R. Hoare, *Structured programming;* Acad. Press (1972) p.1-82.

E.W. Dijkstra, *A discipline of programming;* Prentice-Hall (1976).

D. Gries, *Educating the programmer: notation, proofs and the development of programs;* TR 80-414, Cornell University (1980).

C.A.R. Hoare, *An axiomatic basis for computer programming;* Comm. ACM 12(1969), p.576-581.

Machine arithmetic

W.S. Brown, *A simple but realistic model of floating-point computation;* CS Techn. report no.83, Bell Laboratories (1980).

W.S.Brown & S.I. Feldman, *Environment parameters and basic functions for floating-point computation;* ACM Transactions on Math. Software 6(1980) p.510-523.

T.J. Dekker, *A floating-point technique for extending the available precision;* Num. Math. 18(1971) p.224-242.

T.J. Dekker, *Correctness proofs and machine arithmetic* in L.D. Fosdick (ed.), *Performance evaluation of numerical software*; North-Holland Publ. Co. (1979).

B. Ford, *Parametrization of the environment for transportable numerical software*; ACM TOMS 4(1978) p.100-103.

J.E. Holm, *Floating-point arithmetic and program correctness proofs;* Ph.D. thesis, Cornell University (1980).

D.E. Knuth, *The art of computer programming, vol.2*; Addison-Wesley (1969).

J.H. Wilkinson, *Rounding errors in algebraic processes;* Her Majesty's Stationery Office (1963).

Desirable features of floating-point arithmetic

W.J. Cody, *Analysis of proposals for the floating-point standard;* Computer 14(1981) p.63-68.

J.T. Coonen, *An implementation guide to a proposed standard for floating arithmetic*; Computer 13(1980) p.68-79.

J.T. Coonen, *Underflow and the denormalized numbers*; Computer 14(1981) p.75-87.

T.J. Dekker, *Machine requirements for reliable, portable software* in W. Cowell *(ed.), Portability of numerical software*; Lecture notes in Computer Science 57, Springer-Verlag (1977) p.22-36.

D. Hough, *Applications of the proposed IEEE 754 standard for floating-point arithmetic;* Computer 14(1981), p.70-74.

T.E. Hull, *Desirable floating-point arithmetic and elementary functions for numerical computation* ; SIGNUM Newsletter 14(1979) p.96-99.

C.H. Reinsch, *Principles and preferences for computer arithmetic;* SIGNUM Newsletter 14(1979) p.12-27.

D. Stevenson a.o., *A proposed standard for binary floating-point arithmetic;* Computer 14(1981) p.51-62.

Interval arithmetic

W. Kahan, *A more complete interval arithmetic;* Lecture notes for a summer course at University of Michigan (1968).

S.E. Laveuve, *Definition einer Kahan-Arithmetik* in Nickel (1975) . p.236-245.

R.E. Moore, *Interval analysis;* Prentice-Hall (1966).

R.E. Moore, *Methods and applications of interval analysis;* SIAM Philadelphia (1979).

K. Nickel (ed), *Interval Mathematics;* Lecture notes in Computer Science 29, Springer-Verlag (1975).

N. Apostolatos, U. Kulisch, R. Krawczyk, B. Lortz, K. Nickel & H.-W. Wippermann, *The algorithmic language TRIPLEX-Algol 60;* Num. Math. 11(1968) p.175-180.

F.N. Ris, *Tools for the analysis of interval arithmetic* in Nickel (1975) p.75-98.

J.M. Yohe, *Software for interval arithmetic: a reasonably portable package;* ACM TOMS 5(1979) p.50-63.

Finding a real zero

R.P. Brent, *An algorithm with guaranteed convergence for finding a zero of a function;* Comp. J. 14(1971) p.422-425.

J.C.P. Bus & T.J.Dekker, *Two efficient algorithms with guaranteed convergence for finding a zero of a function;* ACM TOMS 1(1975) p.330-345.

==============

Preparing the NAG Library

B. Ford, J. Bentley, J.J. Du Croz and S.J. Hague

Numerical Algorithms Group, 7 Banbury Road, Oxford OX2 6NN

SUMMARY

If a reliable, high quality numerical algorithm library is to be developed then it is essential that we recognise the need for collaboration between different technical communities in the development of the library. This paper suggests an ultimate design for the library and describes the implications of that design for the people involved in the development of the library.

Key Words Numerical Algorithms Library Library Design Library Development Contribution Validation Assembly Implementation Distribution Library service On-line documentation Machine-based documentation

INTRODUCTION

Since its inception the Numerical Algorithms Group (NAG) project has pursued four aims:

1. To create a balanced, general purpose numerical algorithms library to meet the mathematical and statistical requirements of computer users, in FORTRAN and Algol 60.
2. To support the library with documentation giving advice on problem identification and algorithms selection, and on the use of each routine.
3. To provide a test program library for certification of the library.
4. To implement the library as widely as user demand required.

There are at present 178 members of NAG who are involved in the preparation of the NAG Library. Generally each person has a specific interest or function within the library development process. They may be

contributors, who contribute library contents and write test programs and documentation. They are academics or government scientists who are selected for their individual ability in an area of numerical mathematics.

validators, who certify that the work of the relevant contributor is of the required standard. They are of comparable stature to the contributor in their field of numerical mathematics.

translators, who translate the algorithms into other languages, usually Algol 60 and Algol 68.

implementors, who implement the Library on a particular machine range or one of the 28 full-time staff employed either in the Central Office or as machine range coordinators.

The Group appreciated in 1970 that collaboration between different technical communities, whose members would inevitably be geographically dispersed, was necessary if a library was to be produced. The library would be reliable and of high quality if each phase of the activity was performed to defined standards, in a prescribed manner, in pursuit of specified objectives. A Central Office of full-time staff was established to process, monitor and maintain the contents of the Library and to coordinate the manpower that created and implemented the software.

Two service organisations, one in Europe (Oxford, England) and the other in North America (Chicago, U.S.A.) finally make a Library Service available to users worldwide.

The disciplined employment of the energy, interest and ability of each individual member of NAG in the creation, development and maintenance of the Library is the key to the NAG project. The project is a living flexible organisation - evolving as technical and organisational factors demand.

In the paper we shall suggest an ultimate structure for the Library; outline the major components of the Library activity and review their functions; underline the operational principles of the activity; and finally comment on its performance.

LIBRARY DESIGN

We require a Library structure designed to satisfy the requirements of all users of the Library [1]. The spectrum of users, in their knowledge of numerical analysis, of programming and of problem formulation and solution, will be very broad.

Types of software

In general, however, we can satisfy the majority of their requirements by three types of Library software.

Type	*Function*	*Example*
Problem Solvers:	one routine call to solve the problem	solution of set of simultaneous real linear equations
Primary routine:	each routine contains one major algorithm.	LU factorization
Basic module:	basic numerical utility designed by the chapter contributor for his own and his fellow contributors' use	extended precision inner product

Communication of information

A consistent approach should be employed throughout the Library for communication of information, between its constituent parts and in particular to the user. Wherever possible, information should be passed through calling sequences. The design of calling sequences, the ordering of parameters within them and the naming of routines and variables should be systematic throughout the Library. A common error-mechanism should be used everywhere.

At least three operational requirements can be recognized in the design of each calling sequence:

- convenient and correct use by the programmer,
- satisfaction of the needs of the algorithm,
- use of the data structures of the numerical area.

These requirements underlie the preparation of interfaces for the three types of user software.

- problem solvers: minimum calling sequence,
- primary routines: longer calling sequence, if necessary to permit greater flexibility and control,
- basic module: optimized calling sequence reflecting perceived needs of all contributors yet recognizing demands of efficiency of use.

The three types of Library software could ultimately provide the three tiers of a steady-state library structure. At the present time a number of chapters within the NAG Library do not fit into such a design.

This illustrates a dilemma faced by library designers in general. The state-of-the-art in many branches of numerical software continues to develop, and notions about optimal library design also may change. A library currently in use represents a very significant investment in effort, guided by the state of affairs and of thinking several years ago. The library designer who wishes to keep abreast of developments must not lightly disregard this investment.

Library Documentation

To the programmer, library software is generally only as good as its supporting documentation. The user, at least initially, relies solely on the the documentation to help him find the routine in the library to solve his problem. This involves

- identification of problem type (numerical area: particular numerical characteristics).
- choice of specific routine for problem type.
- advice on the use of the routine.
- study of description of the routine.
- design and development of program calling the routine (perhaps based on the routine example program).

Later he will either turn straight to the specific routine document; or through familiarity with the structure of the library documentation, find the required routine by consulting the keyword index or a particular chapter contents index.

Hence the library documentation consists of three main components; introductory advice and routine selection, description of each routine, and indexes.

The modes of availability of the documentation are also important to the user. Some will only wish, and others only be able to have, access to printed documentation. Full library documentation will be substantial. Hence a mini-manual for introductory advice and routine selection may be useful. But each user must ultimately consult individual routine documents. Microform copy can overcome problems of bulk for transportation and storage. Unfortunately programmers find it less convenient to consult.

Other users, particularly those working through computer networks, will require all documentation to be machine-based and preferably available on-line via visual display units. This simple requirement takes one immediately into the study of character sets, data bases, information systems, operating systems, machine and terminal characteristics and line speeds.

Contents of Library

The contents of the library must evolve as research and development permit. Hence we require a library structure which enables the library contents to change with the minimum inconvenience to users.

It is convenient to divide the contents of the Library in accordance with areas of numerical mathematics. Further subdivision will be required following the natural substructure within each mathematical area. The number of algorithms included will reflect the user demand for problem solution and the resolution of problem type within each area.

NAG LIBRARY PREPARATION

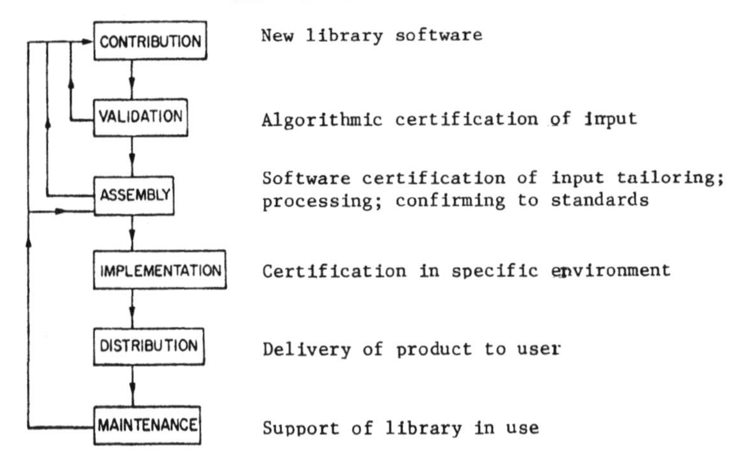

Contribution

The primary function of a contributor is to identify the major types of problems met by users in his area of interest and to provide the 'best' algorithm in the library for each type. The algorithms are selected following a stringent performance evaluation of contending methods. The test program and data sets are retained by the contributor.

Ideally we require routines which will run, virtually without change, to prescribed efficiency and accuracy, on all the machines to which we carry the Library. We need adaptable algorithms [2] which can then be realized as transportable subroutines [3].

The contributor uses an agreed subset of the language [4]. As he has been shown [5], this approach can largely overcome the problem of language dialects, yet permit a reliable and robust subroutine [6] to be developed.

The aim of the contributor is to write a single routine, which can be tailored to perform to the required accuracy and efficiency on each machine.

As a basis for the development of all NAG software, a simple model has been developed of features of any computer that are relevant to the software. This conceptual machine is described in terms of a number of parameters (e.g. base of floating point number representation, overflow threshold). Each parameter may be given a specific value to reflect a feature of a particular computer (e.g. SRADIX [7] is 16 for IBM 360 and is 2 for CDC 7600).

Hence the contributor writes his single routine for the conceptual machine. The routine is then tailored for each distinct configuration by selection of the values of the parameters. In this way the individual demands of accuracy and efficiency for the routine are met for every machine range.

For each routine the contributor provides an implementation test which will be used by implementors to demonstrate the operational efficiency and accuracy of the routine in each environment.

Many users welcome an example of the use of each routine. It is convenient to provide such an example in the user documentation. The contributor writes this example program and provides an example data set, if necessary, together with the results for inclusion in the Library Manual. The contributor prepares draft documentation to support his work. Each routine is then described in an individual document which bears the same name as the routine. The Library Manual has the same chapter structure as the Library. At the head of each chapter is a chapter introduction document which the contributor must also prepare. This document gives background information on the subject area, and recommendations on the choice and use of routines.

Contribution to the Library is an onerous but invaluable activity that makes extensive algorithmic, programming and literary demands.

Validation

The task of the validator is to certify the algorithmic and literary work of the relevant contributor. This second stage in the development of the NAG Library seeks to ensure that the problems addressed by the library contents are relevant to user requirements; that each algorithm is selected after due consideration; and that the user documentation is clear and concise. Substandard or ill-conceived material is returned to the contributor for modification and improvement.

As all these activities involve individual assessment rather than incontrovertible fact; discussion and lively debate often ensue between contributor and validator.

Assembly

Once validated, the software and draft documentation are sent to the Central Office, whose staff are responsible for assembling and processing both the code and the documentation ready for general distribution at each new release (Mark) of the Library.

As the Library is the work of many individuals, there is inevitably inconsistency and confusion in interpretation and in satisfaction of agreed standards. Hence in assembling the codes and documentation provided by contributors, it is essential that we check for compliance with these standards. Wherever possible, these are machine proven, but there are certain standards which need to be checked by hand; for example, we must ensure that the relevant chapter design is being followed and that the user interface chosen for routines satisfies the demands of the general library structure.

Having checked that the draft documentation conforms to content and format standards, the material is added to a documentation data base. The input form is in a type-setting language (TSSD [8]), to permit preparation of photo type-set masters for printing of the Library Manual, but from which an on-line form of documentation is extracted by program [9].

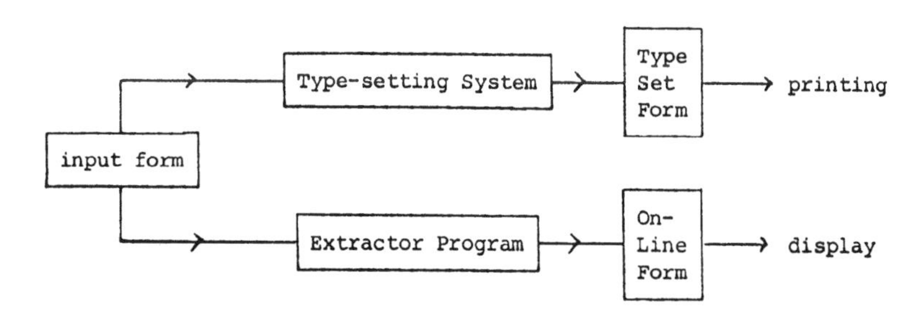

Processing tools available in the Central Office for machine processing of contributed software include the following: [9], [10], [11], [12], [13]

PFORT verifier
DECS
POLISH
BRNANL
APT

Processing of the contributed software by the Central Office, whether automatic or manual, is designed to achieve the following functions:

- diagnosing a coding error either of an algorithmic or linguistic nature;
- altering a structural property of the text, e.g. imposing a particular order on non-executable statements in FORTRAN;
- standardizing the appearance of the text;
- standardizing nomenclature used, e.g. giving the same name to variables having the same function in different program units;
- conducting dynamic analysis of the text, e.g. by planting tracing calls;
- ensuring adherence to declared language standards or subsets thereof;
- changing an operational property of the text, e.g. changing the mode of arithmetic precision;
- coping with arithmetic, dialect and other differences between computing systems.

As further certification of the efficiency, accuracy and effectiveness of the contributed codes, Central Office staff run the routines, together with implementation and example programs, on three different systems (ICL 1900-48 bit floating point word, IBM 370-4/8 bytes and CDC 7600-60 bit word).

Once certified, the routines are used to prepare an updated version of the 'Contributed Library'. This new version, known as a new Mark of the Library, is released once a year and consists of the last generally released version of the Library, supplemented by the newly-certified routines and any other improvements or corrections. For example, at Mark 8, 95 new routines were added to the 395 routines in the Library at Mark 7. The implementation and example program suites, and the NAG Library documentation, are supplemented in a similar manner.

Implementation

One of the tasks of the Central Office is to coordinate all the implementations of the NAG Library. The number of such implementations depends upon precisely what you count; at the crudest level the Library is available on 28 distinct major machine ranges and is being implemented on 2 more. But taking into account

- minor variations in hardware,
- different precision versions,
- different compiler versions;

the numbers of distinct compiled versions of the Library are:

	FORTRAN		ALGOL 60	
Already available	Mk 4	1		
	Mk 5	4	Mk 5	4
	Mk 6	4	Mk 6	1
	Mk 7	37	Mk 7	10
New implementations in train		2		2

There is a good deal of scope for reducing the total amount of work spent in implementation if we take advantage of common features of the various machine ranges. For example, for FORTRAN double precision implementations, a production version of APT (developed in the NAG Central Office) automatically performs most of the conversion from the single precision 'Contributed Library' to a double precision version. This is then the common starting-point for all double precision implementations.

These divide into two subgroups: the 'IBM 360 like machines' with hexadecimal arithmetic and the machines with binary arithmetic.

Implementation tree (FORTRAN)

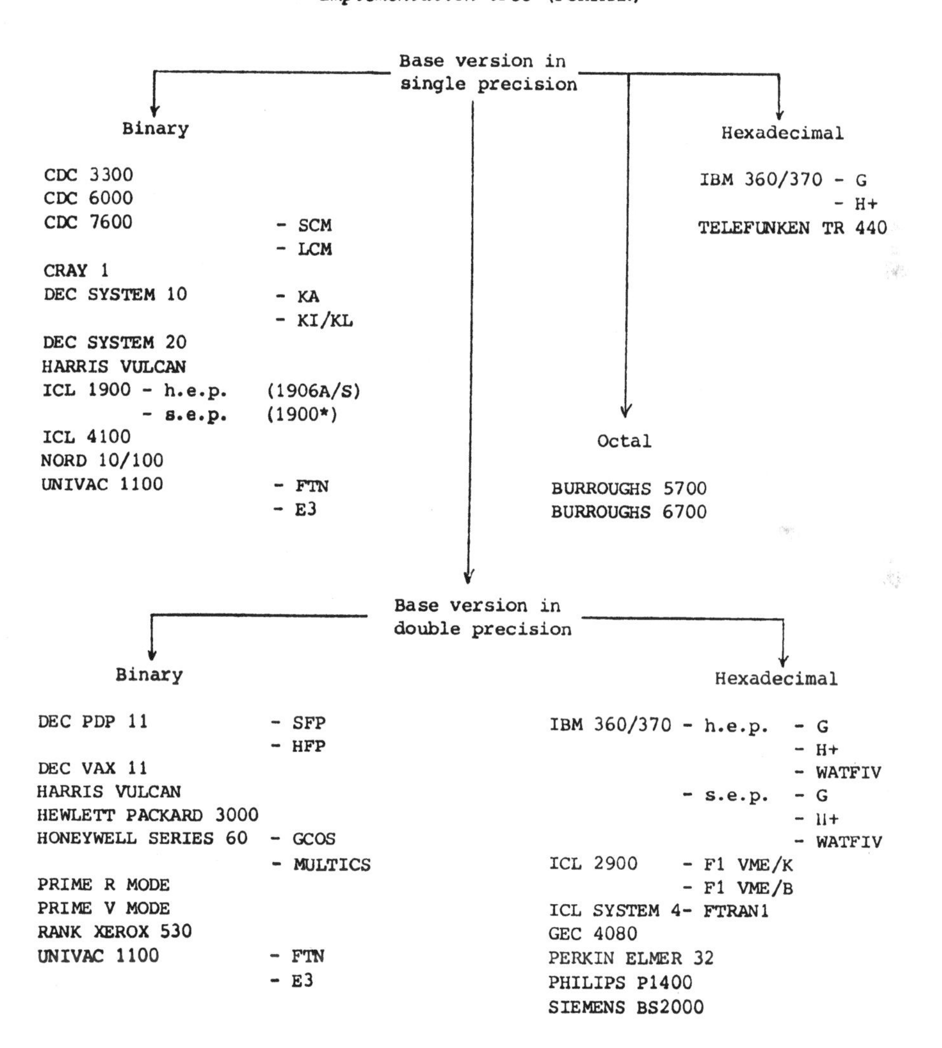

The predicted source texts for each of the machines in the two subgroups have common requirements. All of the byte machines require special modified subroutines in the Special Function chapter. By historic accident, many of the machines with binary arithmetic (e.g. Honeywell, Xerox 530 and Univac 1100 (with FTN compiler)) have FORTRAN compilers which do not include a Double Precision Complex facility. The feature is simulated by Double Precision arrays, with an extra leading dimension of 2, and the amended source text used on these machines.

Coordination of implementations requires the Central Office

- to supply the initial software,
- to advise about any anticipated difficulties (e.g. machine code)
- to help solve any problems which arise,
- to ensure that the implementation is of an acceptable standard (e.g. by examining the computed results).

Each implementation starts from its 'Predicted Library' tape prepared by the Central Office.

The tape holds

- the 'Predicted Library' source text in the relevant precision or precisions,
- the example programs each with their input data and results (all in the relevant precision(s)).
- the implementation programs, each with their input data and results (all in the relevant precision(s)).

The function of the implementor is to read the material into filestore, and then systematically to compile and to test the routines, example programs and implementation test programs. The activity is essentially one of file handling, file management and file comparison. Sophisticated programs have been developed for automatic comparison of results.

Distribution

Each implementor prepares a certified library distribution tape. This contains a precompiled version of the implemented Library, the source text of the routines from which it was prepared, and the example programs plus the input data and results computed during its certification. The structure and format of the tape are chosen to be optimum for the given implementation. The contents and form of the tape are described in a Library Support Note, which also advises the Library site staff how to read the software off the tape into filestore.

Service

The NAG project has always emphasised the importance of distributing, and basing its Library service, on a tested, pre-compiled (load module) Library. Each site simply reads the Library software into filestore, and may then with confidence make it available immediately to their users. There is an annual update of the Library Mark (version) and intermediate correction of software and documentation errors, if required. A numerical and software advisory service by telephone, telex and letter is available from our two offices - and consultancy advice can also be arranged.

OPERATIONAL PRINCIPLES

1. Consultation: enables the Library to address problems met by users, and the NAG Project to serve the needs and requirements of all its parts.
2. Collaboration: gives access to the required expertise and knowledge.
3. Coordination: permits individual parts of the Library to be developed independently, within a unified structure.
4. Planning: permits overall design of individual chapter contents.
5. Standards: ensure the effectiveness and reliability of the product developed.
6. Mechanization: minimizes the cost of development, distribution and maintenance, and is the most reliable method of software processing.
7. Service: encourages wide and regular use of the Library.

CONCLUSIONS

A project such as the one we have described must, of necessity, involve staff with different abilities, background and interests. It has to combine its academic staff, responsible for library contents, and its commercial staff required for servicing the Library. In our organization this involves both volunteers, responsible in the main for providing up-to-date contents and indeed new language versions, and the full-time employees responsible for the co-ordination of the library activity and maintenance of the library service.

For the project to be successful there must be rigour, to ensure that quality and reliability are achieved and maintained, and yet there must be flexibility, to enable cooperation and to optimize the service for each and every implementation. We depend upon enthusiasm to broaden the coverage provided by the Library, and yet we require experience in order to achieve a balanced, general-purpose numerical library service.

REFERENCES

[1] B. Ford and J. Bentley, 'A library design for all parties', in Numerical Software - Needs and Availability (Ed. D.A.H. Jacobs), Academic Press, London, 1978.

[2] B. Ford, S.J. Hague and B.T. Smith, 'Some transformations of numerical software' (in press).

[3] S.J. Hague and B. Ford, 'Portability-Prediction and correction', Software-Practice and Experience, 6, 61-69 (1976).

[4] B.G. Ryder, 'The PFORT verifier', Software-Practice and Experience, 4, 359-377 (1974).

[5] J. Bentley and B. Ford, 'On the enhancement of portability in the NAG project-a statistical survey', in Portability of Numerical Software (Ed. W. Cowell), Berlin, Springer-Verlag, 1977.

[6] W.J. Cody, 'The construction of numerical subroutine libraries', SIAM Review, 16, No.1, 36-46 (1974).

[7] B. Ford, 'Preparing conventions for parameters for transportable numerical software', in Portability of Numerical Software (Ed. W. Cowell), Berlin, Springer-Verlag, 1977.

[8] M.J. Hooper, 'TSSD, a typesetting system for scientific documents', AERE-R 8574, (13), HMSO, London, 1976.

[9] S.J. Hague, S.M. Nugent and B. Ford, 'Computer-based Documentation for the NAG Library', This volume.

[10] J.J. Du Croz, S.J. Hague and J.L. Siemieniuch, 'Aids to portability within the NAG project', in Portability of Numerical Software (Ed. W. Cowell), Berlin, Springer-Verlag, 1977.

REFERENCES

[11] S.J. Hague, 'Software tools', in Numerical Software-Needs and Availability (Ed. D.A.H. Jacobs), Academic Press, London, 1978.

[12] B.G. Ryder, The FORTRAN Verifier: User's Guide, Bell Telephone Laboratories, Technical Report No.12.

[13] J. Dorrenbacher, D. Paddock, D. Wisneski and L.D. Fosdick, 'POLISH, a FORTRAN program to edit FORTRAN programs', Dept. of Computer Science, University of Colorado at Boulder, Ref.: No. CU-CS-050-74 (1974).

[14] L.D. Fosdick, 'BRNANL, a FORTRAN program to identify basic blocks in FORTRAN programs', Dept. of Computer Science, University of Colorado at Boulder, Ref.: No.CM-CS-040-74 (1974).

COMPUTER-BASED DOCUMENTATION FOR THE NAG LIBRARY

By

S.J. Hague, S.M. Nugent and B. Ford

NAG Central Office
7 Banbury Road
Oxford OX2 6NN
United Kingdom

Abstract:

The Numerical Algorithms Group (NAG) develops and distributes a numerical software library in three languages on many machine ranges. We describe the plan by which computerised techniques were introduced for the production of the extensive documentation that supports the NAG Library: in essence the development of a single documentation data base intended for computer type-setting, but from which a form suitable for on-line display can be derived.

Keyphrases:

numerical software library, computer-based documentation, on-line documentation, computer type-setting, Numerical Algorithms Group (NAG), TSSD.

CONTENTS

1. Introduction

Software of even the highest quality is useless without adequate documentation. Providing clear, comprehensive and correct documentation is an essential requirement that software developers must meet, if they wish their products to be widely and properly used. The task of writing such documentation is not an easy one as the generally poor standard of software support manuals shows. It is made harder still if from time to time new facilities (or modifications to existing ones) are introduced, or if the software is adapted for use in different computing environments. *Maintenance* and *adaptation* apply as much to documentation as they do to software: the documentation must 'grow' with the software as it develops and must 'stretch' as it is more widely implemented across machine ranges. (See Appendix B for a list of NAG machine range implementations).

The *Numerical Algorithms Group (NAG)* [1] develops and distributes a mathematical software library in three languages on many machine ranges. In this paper we describe the use of computerised techniques for the production of the existing documentation that supports the NAG Library. These are based upon development of a *single* documentation data base, intended for computer type-setting, but from which a form suitable for on-line display can be derived. This approach is considered in more detail in later sections. First we must describe present NAG documentation to put the work in context.

2. Present NAG Library Documentation

Since its inception in 1970, NAG has aimed to produce extensive, high quality documentation to support its library. Whether or not that quality has been reached is best left for others to judge but that the printed documentation is extensive cannot be in doubt: the five volumes of the Mark 7 FORTRAN Library Manual [2] comprising over 3000 pages testify to that. Each language version of the NAG Library (Algol 60, Algol 68, FORTRAN) is supported by its own Library Manual.

2.1 Manual Structure

The Manual has the same chapter structure as the Library software: chapter E02 is devoted to curve and surface fitting routines for instance. At the head of each chapter there is an introductory document which gives a mathematical background to the problem area e.g. non-linear optimisation, and advises on the choice and use of routines. Such recommendations are especially important where the problems involved are complicated or the roles of different routines overlap. (Advice on which routine to select may take the form of a decision chart in some cases - see Appendix A.2 for an example.) After the *chapter introduction document* comes a list of chapter contents giving a brief summary of the purpose of each routine in the chapter. This list is followed by the *routine documents* themselves, one for each primary (user-callable) routine in the chapter. The routine documents all have these sections:

2.1 Manual Structure (contd)

1. Purpose - a brief statement of the role of the routine
2. Specification - a list of parameters and their associated types
3. Description.- some background to the method (algorithm) used
4. References - further reading
5. Parameters - a description of the use of each parameter of the routine
6. Error Indicators - error messages generated by the routine
7. Auxiliary Routines - other NAG Library routines used
8. Timing - usually symbolic expression, e.g. 'varies as N^3'
9. Storage - as section 8
10. Accuracy - expected bounds on computed solution
11. Further Comments - additional information usually of a programming nature
12. Keywords - for a keyword data base
13. Example - a complete program with data and results, showing how to use the routine.

Appendix A gives an example of a typical printed FORTRAN Library routine document.

2.2 Updating to New Marks

As we stated earlier, documentation must keep up with software both as it develops in terms of facilities and as it becomes more widely implemented. The updating problem is eased by producing the Library Manual in a loose-leaf document-based form which permits the addition of new documents (or pages) and the replacement of existing ones. However, introducing a new routine, with its 10-sided (say) routine document, is not simply a matter of addition. Indices in various places have to be modified as do at least those parts of the appropriate chapter introduction which may now have to advise on the use of the new routine in place of or in conjunction with other routines. To give an idea of the scale, consider the Mark 6 to 7 update of the FORTRAN Library Manual. The update involved about 1100 pages, of which 900 or so pages were specific documentation for the 80 new routines added at Mark (edition) 7. About 130 of the remaining pages replaced existing pages to give updated lists and advice, and the rest removed defects of one kind or another. The net increase in size between Mark 6 and Mark 7 was about 800 pages because some routines (and their associated documents) were withdrawn at Mark 7.

2.3 Adapting to New Environments

The design problem of documenting an expanding body of software is primarily a structural one: how to insert new material or replace existing sections. Coping with new computing environments is more a matter of judgement in technical content and of confidence in the behaviour of the software on different machines. Implicit in the idea of *transportable numerical software* [3] is the belief that the effect of arithmetic and other differences between computing systems can largely be foreseen, particularly as more experience of multi-machine implementation is gained. Thus documentation can be written in a slightly generalised manner where necessary so that critical information, e.g. the minimum tolerance for convergence, can be readily interpreted by the user in his own computing environment.

2.3 Adapting to New Environments (contd)

NAG republished its Algol 60 and FORTRAN Library Manual in this machine-independent form in 1975, and so far the aim of having each language version support all implementations in that language appears to be successful. (In any case, the alternative of having a Library Manual per implementation would be economically impossible.) Some global information about the use of the Library in a particular environment may be necessary (e.g. precision in FORTRAN) and any specific exceptions must be noted. For instance, test results produced during the implementation process may diverge by an unexpectedly large amount from supplied reference results.

3. Why Computerise?

Having completed our description of the NAG Library Manual, we now consider the perceived advantages of computerisation that caused us to replace our conventional techniques of producing master documents on IBM 'golf-ball' typewriters. Prior to Mark 7, the only form of computerised information provided by NAG was a summary of the purpose of each routine. What factors made us contemplate a wholesale move to computer-based documentation?

- the ever-growing size of the Library Manual made it increasingly unwieldy. More compact printed forms were sought and alternative media (microfiche perhaps) explored.

- there is only one set of typed masters and some of those masters were over 3 years old (the FORTRAN and Algol 60 Manuals were re-typed in 1975/76). The older masters particularly are heavily amended in some cases and further modification would necessitate re-typing.

- little scope existed for producing derived publications (other than 'pure' subset publications like the Mini Manual) because there is just one set of ageing masters.

- there are not many technical typists who are prepared to produce high quality, detailed work over a long period of time.

The advantages of computerising the preparation of master documents are obvious:

- the machine-readable masters are *re-producible*.

- they are easier to *update*.

- there is greater scope for producing variant publications by *transforming* or selecting parts of the master document files.

3. Why Computerise? (contd)

A move to computer-based techniques appeared highly desirable but there were difficulties to be faced. Changes in production technology can pose personnel and organisational problems. Appropriate computing facilities (both hardware and software) must be available - the unsuitability of the ICL 1900s for text processing (the NAG Central Office has access to the ICL 1906A of Oxford University Computing Service) has been a major reason why NAG did not computerise its documentation from the beginning. These aspects of resources and equipment would apply to any organisation which is contemplating such a move. Two further, NAG-specific, constraints that we had to satisfy were:

- it must be possible to list Library documentation, possibly in some 'reduced' state, on a computer terminal or line printer, *but*

- printed documentation derived from computer-held masters should in no significant way be inferior in appearance to the earlier conventionally produced material. New style documentation must be no less attractive than the old, and must place no major constraint on the 'freedom of expression', particularly in the mathematical sense. Ideally it would be preferable in appearance to the earlier material.

Our plan to satisfy these conflicting constraints and the other requirements mentioned above is described in the next section.

4. Computerisation - The Basic Plan

Type-setting offers the prospect of fine control over layout, a wide range of characters and different sizes (and boldness) of script. An important element of NAG Library documentation is the use of mathematical expressions, so any *computer-based type-setting system* under consideration must provide sufficiently powerful features for composing such expressions. It must also have adequate facilities for general text processing e.g. vertical and horizontal spacing, indentation control, headings, footings, justification. Given such a system, it should be possible to maintain or to improve upon the quality of the conventionally produced masters. The data presented to a type-setting program would be Library documentation prepared in a computer type-setting language. To meet the other constraint stated at the end of Section 3, however, a much simpler data base seems necessary. Listing documentation at a terminal or printer means that a far more restrictive character set must be employed and minimal assumptions about carriage control made. This is particularly so since the NAG Library is widely implemented, and such documentation would be listed on a wide variety of printing or display devices. To avoid having two separate data bases, the essence of our approach is to have a single data base, prepared in a type-setting language, but from which the *on-line form is extracted* by program:

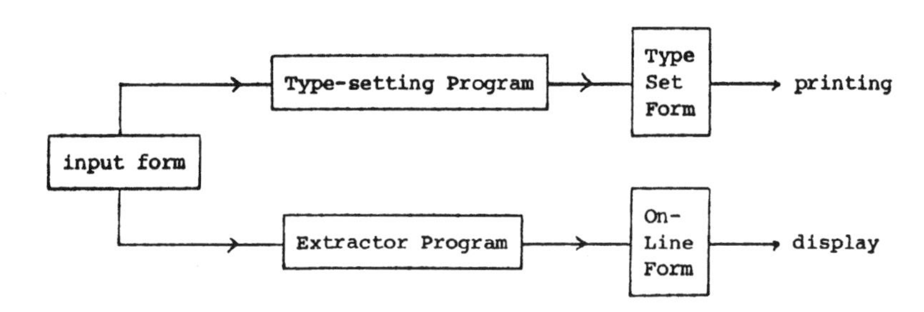

The computer type-setting system used, TSSD, is described in the next section, and this is followed by a description of Cuneiform, the present version of the *extractor program* developed for use with a TSSD-compatible data base.

4.1 A Type-setting System for Scientific Documents (TSSD)

TSSD is a computer type-setting system which can be used to type-set scientific documents in preparation for printing. It has been developed by M.J. Hopper of AERE, Harwell. A description of the initial system mounted on the Harwell IBM 370/168 is contained in [4], issued in November 1976. Since that time further facilities have been incorporated ([5]). The TSSD program is written mostly in FORTRAN and produces type-setting instructions for the Photon Pacesetter mark I photo-type-setter at Harwell. We now describe the major features of TSSD.

(a) Design aims

The principle reason for developing TSSD was to assist in the documentation of the Harwell Subroutine Library. That Library is supported by extensive documentation, much of which is of a similar nature to that in NAG Library manuals. The arguments for computer type-setting advanced in Section 4 are therefore echoing the views expressed in the original Harwell report about the motives behind TSSD. If it was to be successful, it had to be able to exploit the full ranges of type-setting facilities to achieve "the clarity of presentation necessary for our users to extract the information quickly and without ambiguity. Allied to this general aim was the requirement to handle mathematical expressions of arbitrary complexity. Thus the second aim of TSSD was to provide a powerful mathematics type-setting facility which was simple to use and which could relieve the user of type face selection, size selection or positioning. Whilst providing as much type-setting power as possible, TSSD was also designed to be easy to use. Input to TSSD should be reasonably readable but, more importantly, the user interface to the system would ideally be only as sophisticated as the application demanded. To achieve this, a '*define*' facility was introduced to allow users to design macros which collectively could form a new type-setting language tailored to their needs. (See Section 6 of [4] for an example).

4.1 A Type-setting System for Scientific Documents (TSSD) (contd)

Before Harwell began to develop TSSD other systems were examined. All were found lacking in one respect or another, or were too embedded in one specific computing system. TSSD has itself been assessed by NAG. We have found transfer of the system to another computing configuration feasible (if not straightforward), and as a language, its general type-setting power adequate for our needs, and the mathematical facilities are truly impressive, an assertion supported by the example given in Appendix D (reproduced from [4], by kind permission of the United Kingdom Atomic Energy Authority).

(b) Language structure and commands

Input to TSSD consists of a set of typed lines, each not more than 72 characters in length. The characters belong to the ASCII set. Each line is divided into words where the word boundaries are recognised by the occurence of any one of several separator characters. Certain characters have special functions such as heralding a TSSD command or coded symbol representation.

A TSSD *input word* can consist of normal word (possibly with coded character representations), a literal string, a TSSD type-setting command, an operand of a command, a single special character, or a define reference (macro call).

There are two modes of type-setting: normal and mathematical. Because general text processing and mathematics pose distinctly different type-setting problems, they each have their own sub-language within TSSD, though it is easy to switch from the normal mode to the maths mode and vice versa.

For a summary of maths mode type-setting, see (c) below

Commands in TSSD take the form

$<two character command name><operands>

e.g.

$TF 4 - select type face 4 , or
$LL 6.5i - set line length to 6.5 inches.

For some commands, the <operands> part may be null, for others it may be an expression or a list of elements, which might be expressions. These expressions may include the common algebraic operators and parentheses, and refer to the values of tabs, registers or mark points. TSSD allows the use of up to 40 tab points, 100 register values and 10 mark points. Reference to the value associated with a basic command can now also be made:

e.g.

$LL 6i - sets line length to 6 inches, and
$VS LL/10 - sets vertical spacing between lines to a tenth of the current line length.

A major feature of TSSD is the availability of a *define string* ($DS) command. This allows the user to define his own macros which expand into basic TSSD commands. Paramteres can be passed from the calling level. The format of the command is

$DS *text* <delimiter> <string not containing delimiter> <delimiter>

The first operand *text* is the *define reference* (or macro name) and the second operand specifies the substitution action.

In normal mode, define references are preceded by a * , thus

$DS N '$NL' means that *N, instead of or as well as $NL, can be used thereafter to introduce a new line. Define string commands can also be used in maths mode but the two sets of commands are kept separate.

The second operand of $DS, the *defined string*, can contain any character or command, including other define references (but recursion is not permitted). A defined string can also fetch words from outside the string by the use of a * at the point where the TSSD word is to be included, e.g.

$DS FACE '$TF *' is

equivalent to $TF 3 when

*FACE 3

occurs in the text. There can be more than one fetch in a string and they can be nested to several levels.

Four commands exist for conditional action. The first has the form

$CE *exp1 op exp2*

where op is a comparison operator. The other three are $TC for action when $CE yields the value true, $SF for the false case, and $EC for ending the conditional action. By using the define string facility to refer to $CE as *IF, $TC as *THEN etc, a structured form of conditional expression can be used:

*IF *exp1 op exp2*
   *THEN *true clause*
   *ELSE *false clause*
*ENDIF

(c) Summary of normal mode facilities

The facilities for type-setting general text include:

- type face selection
- type size selection
- line lengths and indents
- justification and hyphenation
- forcing new lines
- vertical spacing between lines
- forward spacing within lines
- other spatial movements within line
- alignments and break-points
- gap-filling (leaders),

and, more recently introduced,

- page control (width, forced new page, numbering)
- column control (multi-column work, balancing, justification, forced new column)
- folios, headers and footers (for text at top/bottom of page)
- protected sections or 'keeps' (not to be split)
- interrupt of processing after *n* lines
- printing of expressions evaluated during processing
- proofing option for intermediate form listed to line printer
- reverse tab switch.

There are also two commands for switching to and from maths mode ($SM and $EM, alias [and]) and a third command ($ML) to alter permanently the spacing between lines when placing one equation above another.

(d) Mathematical mode

The intention in this mode is that the user sets an expression by presenting to TSSD a verbal statement of what the expression represents mathematically, for example

$$\int_0^1 \frac{e^{-\lambda t}}{1+t} \, dt$$

requires the input statement

```
[INTEGRAL FROM 0 TO 1 [e SUP[- lamda t]] OVER 1+t_dt]
```

Positioning, size changing and typeface selection are all done automatically by TSSD using printing industry conventions. By default, Roman letters appear in italic form, and numerals in upright medium script. Apart from this, when maths mode is entered all type-setting values and settings remain operative. On leaving maths mode the situation reverts back to the state on entry. Changes made in maths mode have no affect on the status of normal mode.

Maths mode permits the use of define references which may either be a simple replacement of text by text or an exchange of text by a more complicated string of keywords, text and other define references. The use of *lamda* in the above example is an instance of the simple case: the greek symbol λ is substituted.

Each character is considered to reside in a *box*. As the parts of a mathematical expression are processed, boxes are joined together to produce enlarged boxes, thus

box OVER *box* → *box*

1 + x OVER 1 + y → $\frac{1+x}{1+y}$

The height, width and position of the newly-produced box are adjusted according to context.

4.1 A Type-setting System for Scientific Documents (TSSD) (contd)

In summary, the facilities available in maths mode are:

- typeface selection, e.g. ROMAN or BOLD
- superiors and inferiors e.g.

  a SUP 2t → a^{2t}

- square toots, SQRT *box*
- fractional expressions, OVER
- TO and FROM, e.g.

  @CA FROM i=1 TO n → $\sum_{i=1}^{n}$
  ↑
  (direct reference to summation symbol)
- selecting size, SIZE, applying only to box that follows it. Also LEFT *text box* , which matches the height of the text item to that of the box, and similarly

  box RIGHT *text*
- positioning equations above equations, e.g.

  PILE [*box* ABOVE *box*]

  There are associated commands LPILE,CPILE,RPILE and TPILE for piling aligned equations on the left, in the centre, on the right and according to a specified tab.
- matrix alignment, i.e. row <u>and</u> column alignment,

  MATRIX[*c*COL[*box* ABOVE *box*...ABOVE *box*]*c*COL[...]...]

  where *c*COL can be any of COL, LCOL, CCOL, RCOL, TCOL which have similar meanings to the corresponding PILE commands.

Finally we must remark on the complexity of TSSD maths mode. It is extremely difficult to devise a simple means of describing mathematical expressions which are often typographically complex. All the operators, some of which are mentioned above, have precedence values just as the more familiar arithmetic operators do, e.g. we would normally interpret

  x*y-z as (x*y)-z , that is,

multiplication 'binds' more than subtraction. One difficulty in understanding TSSD maths mode is that to the beginner, the operators and their precendence values are unfamiliar. The use of square brackets [,] to obtain the correct grouping of elements and to conform to syntax rules is crucial. All maths mode commands can be nested to any level.

(e) Application to NAG documentation

In preparing NAG routine documents we have sought an optimal compromise between clarity and compactness. A standard set of NAG-oriented type-setting commands quickly emerge. By making extensive use of the define string ($DS) basic command mentioned earlier, the NAG set of commands are defined to meet the specific needs of NAG routine documents. This set of command definitions have been automatically inserted in every routine document file. The following examples illustrate the way in which the use of these commands simplify the preparation of computerised NAG routine documents:

.....*BOLD 'D01AGF'.....
- causes the string 'D01AGF', a NAG routine name, to be type-set in a bold script. It is defined as

  $DS BOLD '$TF4' , i.e. as Type Face 4.

.....*LSPACE.....
- causes a new line, drops vertical spacing by 7 units, sets left indent to current left indent plus 5 space units. It is defined as

  $DS LSPACE '$NL *SGAP $IL I+5n'

  where $NL and $IL are basic commands for forcing a new line and setting left indents, and *SGAP is defined as

  $DS SGAP '$VS 7'

  where $VS controls vertical spacing.

.....*HEAD1.....
- inserts a section heading '1.Purpose' in bold script. Defined as

  $DS HEAD1 '*BOLD 1. Purpose'

More complicated commands, *HEADER and *FOOTER, are designed to give information to appear at the head and foot of each page. Other commands control the use of layout, section headings elsewhere in the document and standard warnings to be inserted at certain points. Defining all these commands once and then referencing them in the body of the document does not just simplify and shorten the task of data preparation and input. It allows us to see what the effect of modifying a command would be throughout a document (or set of documents). This would not be the case if only basic TSSD commands were incorporated in full in *each* place in the document where some particular effect was required (e.g. indenting all parameter descriptions by a specified number of space units).

4.2 Extractor Program for On-line Form

(a) Design aims

The role of a type-setting program such as TSSD is to assist in the composition of high quality master documents, from which the printed Library Manual is produced. The printed form will remain the basic source of reference for the majority of users of the Library.

4.2 Extractor Program for On-line Form (contd)

There is, however, demand from some sites for an intermediate computerised form of reference documentation which can be listed by users at terminals or on line-printers. This intermediate or *on-line form* has been derived from the TSSD-compatible documentation data base and is a *reduced form* in at least two senses:

(i) it must be a simpler, more *restrictive representation* of the information in the Library Manual because it has to be capable of reproduction on a wide variety of devices including terminals with short line widths, no back space or underline facility and little more than the FORTRAN-66 character set available.

(ii) some parts of the documentation, particularly mathematical expressions, are very difficult to express clearly on a conventional terminal or printer. Moreover the slow speed of some devices, the limited storage available at some user sites for computer-held documents, and other factors concerned with timescales all indicate that the on-line form must have a *reduced information content:* that is, not all 13 sections of each routine document will necessarily be provided.

Reduction (i) is a technical problem solved by writing an extraction program which maps the type-setting language into a simpler form. Elements of that mapping process are described in the next section. The second reduction is more philosophical in nature because we must ask ourselves what sections of the routine document can we afford (for the user's sake) to leave out. What is to be the role of the on-line form and how will it relate to the printed reference documentation? These are important questions that exercised our minds for some time.

We concluded that the on-line form should provide programming information (as opposed to algorithmic details) about a routine. Thus the sections to be provided would include those describing the purpose of the routine, the types and purpose of parameters, and error indicators, i.e. Sections 1,2,5 and 6 of the routine document. For some routines, part or all of other sections might be included in a combined Further Comments section. (The example program, data and results are provided in machine-readable form to user sites and could be presented in a combined way with the programming information.) However the core of the on-line form is those four sections, re-numbered as A, B, C and D. It must be admitted that the Library Manual was not consciously designed with this separation in mind, and occasionally some re-casting of the total information content is necessary to make the extracted sections stand alone to form an extended *summary of use* which should serve as:

- a detailed pointer to the suitability of a routine for a user's problem. The user then consults the printed publication.
- a reminder to the user about the role of a particular parameter or about the significance of an error indicator, where the user has already consulted the full documentation.
- adequate, self-contained documentation of a routine either where the routine itself is straightforward to use e.g. sorting, or where the user is familiar with the subject area and understands the concepts and terms used.

4.2 Extractor Program for On-line Form (contd)

In all cases, the on-line summaries of use inform the reader that more extensive documentation is available in the printed Library Manual and for some of the more sophisticated or obscure routines, that warning is reinforced. An additional, potential role for the on-line form is concerned with the adaptation issue discussed earlier. We noted that to have one Library Manual support all implementations meant that some generality of expression had to be introduced. If the on-line were to be distributed to user sites via the implementation framework, this would give an opportunity to remove some of that generality which will persist in the typeset printed form. For instance, IBM double precision users need not be told that a parameter is of type *complex* or //COMPLEX//, where the interpretation of such terms is given elsewhere, but is indeed of type COMPLEX*16, whereas the ICL 2900 double precision user could be informed that the type is COMPLEX*D. The extent to which the on-line form should or could be tailored to implementation or even user - preference is discussed further in Section 5 of this paper.

The general task of the on-line extractor program, currently called *Cuneiform*, is then to extract Sections 1,2,5 and 6 from routine documents in TSSD form. Its main job is to recognise TSSD commands and interpret them appropriately for the on-line form. For some commands the appropriate action is to do nothing but ignore them but it must nevertheless still recognise even these commands. The stability of the TSSD data base is therefore of crucial importance: when a new command is introduced, Cuneiform must be extended to cater for it. However for commands that are already recognised there is still considerable scope for experimentation, particularly regarding type-setting. For instance the *CAPITAL command provides document heading information for both TSSD and Cuneiform but the fine detail of the TSSD definition of *CAPITAL (using the ubiquitous $DS define string) could be adjusted without disturbing the on-line extraction process.

(b) Options and interpretation of commands

Cuneiform takes as input routine documents in TSSD-compatible form. It has the same overall rules as TSSD for recognising words. The commands it recognises are of the form *<command>, which have been defined elsewhere for type-setting purposes, using the $DS basic command. To illustrate the application of the extractor program we give examples below of some of the commands and their *dual interpretation:*

Heading commands: *CAPITAL,*HEADn,*SECTION, ...

e.g.

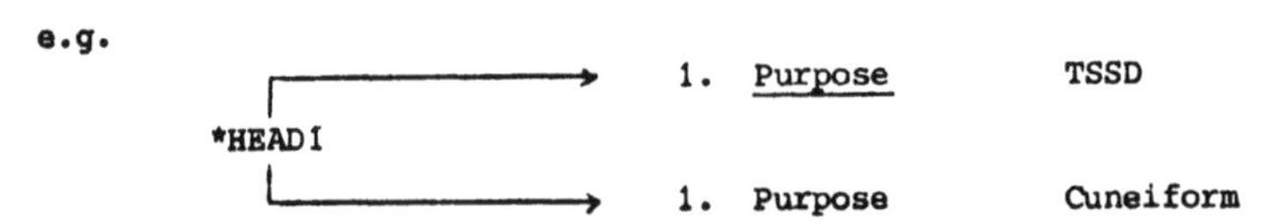

4.2 Extractor Program for On-line Form (contd)

Text commands: *BOLD,*ITALIC,*SPECIAL,*MEDIUM,*TYPE,*WARNING,*NAG, ...

e.g.

*ITALIC → *real* (TSSD)
*ITALIC → **real** (Cuneiform)

*WARNING → inserts standard warning in on-line form but not in type-set version

*TYPE → restores typeface in type-set version but has no effect on Cuneiform

*NAG F01AAF → F01AAF (TSSD)
*NAG F01AAF → //F01AA?// (Cuneiform)

where ? is F by default but could be set to any other character (E or D for instance) by using *LAST command.

Layout commands: *LSPACE,*N, ...

e.g.

*LSPACE → causes two extra leading spaces to be inserted on current and subsequent lines.

Mathematics commands: *GE,*OD,*PD,*MU, ...

e.g.

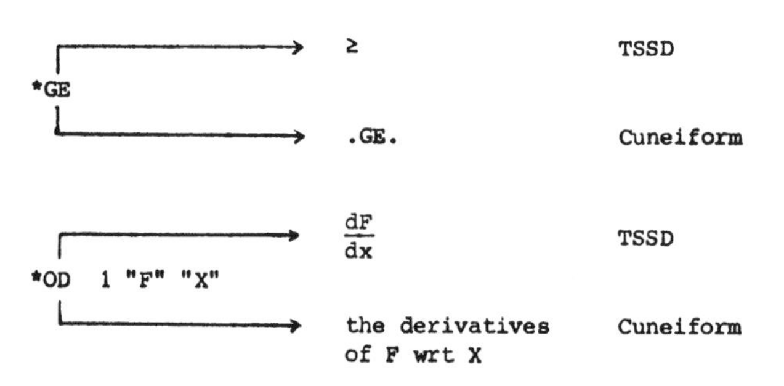

4.2 Extractor Program for On-line Form (contd)

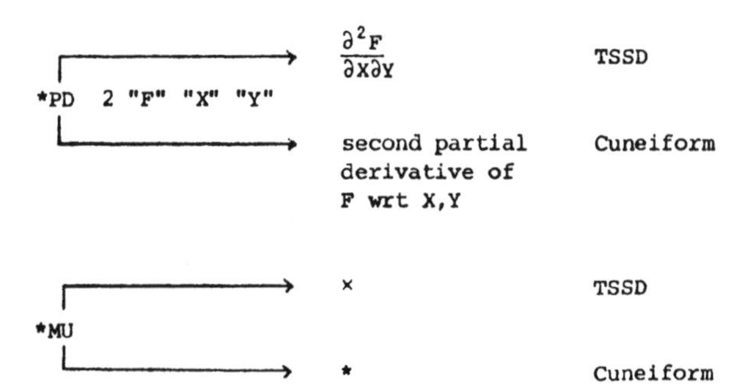

The above commands provide a partial answer to the representation of some concise, standard mathematical devices. Provision must be made for the more general case. TSSD mathematics mode is introduced by an opening square bracket [and terminated by a closing square bracket]. Cuneiform ignores matching square brackets and their contents. Alternative mathematical text can be held in \...\ delimiters. Usually the brackets and \...\ are adjacent but need not be.

e.g.

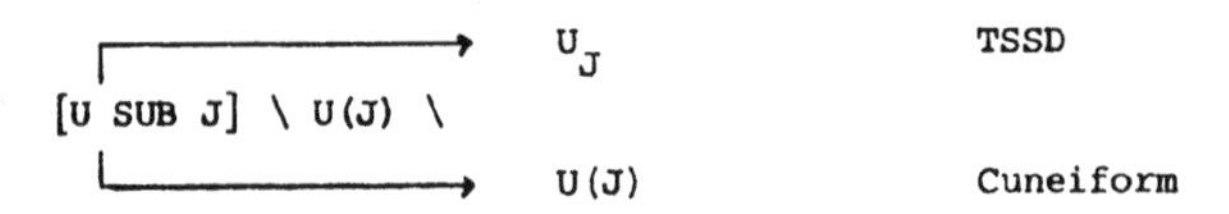

This alternative expression facility is appropriate for compact expressions but for more extensive mathematical statements it is better to use the general comment facility ....

Comment commands: *SKIP,*C ...

e.g.

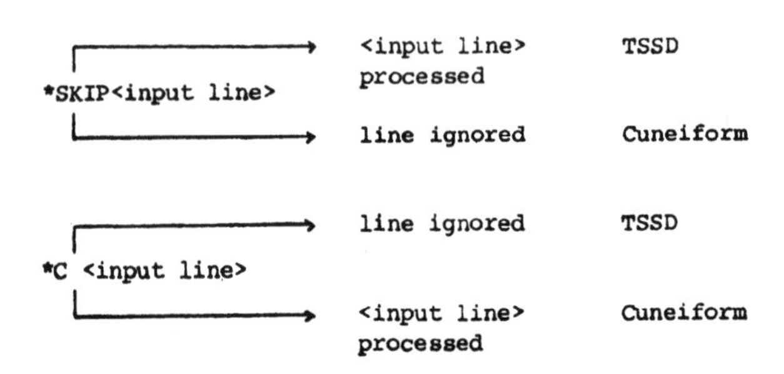

Extractor Program for On-line Form (contd)

Formatting options:

Cuneiform provides the following facilities that can be adjusted by the appropriate commands -

- automatic pagination after n lines of output
- adjustable line length
- left and right justification
- single or double column listing options.

Character mappings

The character set into which that used in the TSSD base is mapped can be adjusted by modifying a look-up table. The present default mapping is into the FORTRAN 66 character set plus lower case letters and colon.

5. Implementation of the Plan

The Cuneiform extractor program [6] has been implemented in FORTRAN on the ICL 2980 of Oxford University Computing Service. All routine documents in the linear algebra chapters of the Mark 7 FORTRAN Library Manual have been brought to the on-line level on the 2980: that is, at least Sections 1,2,5 and 6 have been input in TSSD form and then processed by Cuneiform. The output, after checking, was transmitted via Library machine range implementors to user sites as an experimental facility at Mark 7.

At Mark 8 all the FORTRAN routine documents are at the on-line level, and henceforth this complete coverage will be maintained. In the light of the experience gained between Marks 7 and 8, we have a more developed opinion about *decentralisation* and *specialisation*.

decentralisation: the information distributed at Mark 7 is the output of an extractor program applied to a data base held *centrally*. That program has options to control pagination, justification, line width etc. In the initial venture, all these options are to be set by the Central Office, and therefore the same extracted information is being distributed via implementors. By providing the data base and an extractor program to implementors or even user sites, however, the setting of these and other options which may be introduced could be decentralised.

specialisation: one option in Cuneiform permits the last character of a NAG routine name to be changed to some other character. This facility could be useful at a site where both single and double precision implementations of the NAG Library are available since the two precision versions of the same routine would differ in the sixth character position (e.g. A02AAE - single, A02AAF - double). Another way in which the generalised information in the printed documentation could be made more implementation-specific in the on-line form is by substituting the actual precision type (be it REAL or DOUBLE PRECISION) for the symbolic precision type, *real* or //REAL//.

5. Implementation of the Plan (contd)

> As well as making such global changes, the implementor could introduce more specific revisions e.g. by substituting actual values, instead of pointers to values stated elsewhere, in mathematical and machine constant chapters. User sites might also want to embed information into the on-line form where local options, e.g. the logical channel for output of error messages, have to be set.

Both these trends seem in principle desirable but they inevitably raise questions of the control and validity of information.

As for TSSD itself, several technical and logistic queries have been overcome, but others remain. We have been able to implement the system in a non-IBM, but byte, environment and to adapt it to another range of type-setting equipment. Can we move the system to a bit machine, and make to other type-setters a limited, straightforward activity?
A further possibility that we must ultimately face is adopting a type-setting system other than or as well as TSSD at some later stage. In such an event we would presumably attempt to transform the existing data base mechanically and modify Cuneiform or its successor where necessary.

Having successfully implemented TSSD (Mark 3) on an ICL 2980 and interfaced it to an APS type-setter, we have used it to produce all new routine documents at Mark 8 in computer type-set form. These documents (92 in all) and other conventionally-typed pages form the Mark 7 to 8 update of the FORTRAN Library Manual. More type-set material could also be phased in at later Marks. After a period of several years the entire manual would be printed from computer type-set masters. We must stress, however, that the phasing-in plan is speculative at this stage: several factors, not least the one of finance, will decide its fate. If and when we might adopt a similar policy towards other major NAG publications such as the Algol 60 or Algol 68 Manuals is also open to question but the same technical strategy could be followed.

To meet our Mark 8 FORTRAN objective required a substantial proportion of our resources, having surmounted many preliminary obstacles. The prospect of an entirely computerised manual (or at least computerised in substantial, contiguous parts) within a few more years will only be realised with a sustained effort. If all that effort was just for the sake of technical vanity or simply to make it easier to correct or modify text, it would certainly be highly questionable. The scale of manpower, equipment and computing resources involved should not be under-estimated. We believe that the effort is justified, however, when the longer term benefits and possibilities are considered as well. For instance, we have concentrated in this paper on NAG routine documents but information at the chapter and library level is included in out plan. Computerising the decision trees in chapter introduction documents requires particular attention but also begs the question: should the computer-held form simply be a machine-readable image of the present, conventionally typed form? In the longer term, it ought to be more than just that. We look forward to the day when the user can interrogate those trees in an interactive manner.

5. Implementation of the Plan (contd)

Similarly we started at Mark 7 to provide keywords with the on-line form but only as they appeared in the existing printed manual. We know of several user sites that plan to embed these keywords in their own local 'help' systems but at some stage, NAG should provide a portable interrogation utility for use with such an index of keywords and phrases. This would be just one more step towards providing a more complete *information service* than we do at present.

Another possibility presented by computerisation is that of sub-setting or re-grouping of documentation to reflect changes that might one day be made to the structure and presentational form of the Library. More speculative but still worth considering is the prospect of documentation translation. Suppose that NAG decided to develop a language-X Library by mechanical translation from one of the existing language versions. The translated code might be acceptable only as a first approximation to the final product but might not a first attempt at language-X user reference documentation be also produced mechanically?

6. Conclusion

The plan has been implemented and appears to be successful. User reaction to the products of the development has yet to be experienced. This will be, of course, the ultimate test.

7. Acknowledgements

In their various ways many people have contributed towards present NAG documentation either by helping to design, write, check or produce it. Some of those people have for some time been firm advocates of computerisation, though they recognised some of the difficulties that its introduction would entail. That we have come as far as we have is in part a tribute to their persistence. These individuals include Dr. A.A. Hock of Leeds University and Dr. D.B. Taylor of Edinburgh University. Our TSSD mentor has, not surprisingly been Mr. M.J. Hopper of AERE, Harwell.

8. References

[1] FORD, B., BENTLEY, J., DU CROZ, J.J. & HAGUE, S.J., "Preparing the NAG Library", This Volume.

[2] NAG FORTRAN Library Manual, Mark 7. NAG Central Office, Oxford, 1979.

[3] HAGUE, S.J. & FORD, B. "Portability - Prediction and Correction", Software Practice and Experience, Volume 6, No.1, January, 1976.

[4] HOPPER, M.J. "TSSD, a typesetting system for scientific documents", AERE-R 8574, (13), HMSO, London, 1976.

[5] HOPPER, M.J. "Enhancements to TSSD", Harwell internal note, May, 1978.

[6] NUGENT, S.M. "The Cuneiform Extractor Program", NAG Internal Note, NAG Central Office, Oxford, 1979.

APPENDIX A1

G08AFF – NAG FORTRAN Library Routine Document

NOTE: before using this routine, please read the appropriate implementation document to check the interpretation of ***bold italicised*** terms and other implementation-dependent details. The routine name may be precision-dependent.

1. Purpose

G08AFF performs the Kruskal–Wallis one-way analysis of variance by ranks on k independent samples of possibly unequal sizes.

2. Specification

```
      SUBROUTINE G08AFF (X, L, LX, K, W, H, P, IFAIL)
C     INTEGER L, LX(K), K, IFAIL
C     real X(L), W(L), H, P
```

---

3. Description

The Kruskal–Wallis test investigates the differences between scores from k independent samples of unequal sizes, the i(th) sample containing l_i observations. The hypothesis under test, H_0, often called the null hypothesis, is that the samples come from the same population, and this is to be tested against the alternative hypothesis H_1 that they come from different populations.

The test proceeds as follows:

1. The pooled sample of all the observations is ranked. Average ranks are assigned to tied scores.
2. The ranks of the observations in each sample are summed, to give the rank sums R_i, $i = 1,...,k$.
3. Kruskal–Wallis' test statistic H is computed as:

$$H = \frac{12}{L(L+1)} \sum_{i=1}^{k} \frac{R_i^2}{l_i} - 3(L+1),$$

where

$$L = \sum_{i=1}^{k} l_i,$$

i.e. the total number of observations. If there are tied scores, H is corrected by dividing by:

$$1 - \frac{\Sigma(t^3 - t)}{L^3 - L}$$

where t is the number of tied scores in a group and the summation is over all tied groups.

G08AFF returns the value of H, and also an approximation, p, to the probability of a value of at least H being observed, H_0 is true. (H approximately follows a χ^2_{k-1} distribution). H_0 is rejected by a test of chosen size α if $p < \alpha$. The approximation p is acceptable unless $k = 3$ and l_1, l_2 or $l_3 \leq 5$ (in which case table O of [1] should be consulted) or $k = 2$ (in which case the Median test (see G08ACF) or the Mann–Whitney U test (see G08ADF) is more appropriate).

4. References

[1] SIEGEL, S.
Nonparametric Statistics for the Behavioral Sciences, Chapter 8.
McGraw Hill, 1956.

5. Parameters

X – ***real*** array of DIMENSION at least (L).

Before entry, the elements of X must contain the observations in the K groups. The first l_1 elements must contain the scores in the first group, the next l_2 those in the second group, and so on.

Unchanged on exit.

L – INTEGER.

On entry, L must specify the total number of observations.

Unchanged on exit.

LX – INTEGER array of DIMENSION at least (K).

Before entry, LX(i) must contain the number of observations l_i in sample i, for $i = 1,...,K$.

$$L = \sum_{i=1}^{k} LX(i).$$

Unchanged on exit.

K – INTEGER.

On entry, K must specify the number of samples , k.
$K \geq 2$.

Unchanged on exit.

W – ***real*** array of DIMENSION at least (L).

Used as workspace.

H – ***real.***

On exit, H contains the value of the Kruskal–Wallis test statistic.

P – ***real.***

On exit, P contains the approximate significance , p, of the Kruskal–Wallis test statistic.

IFAIL – INTEGER.

Before entry, IFAIL must be assigned a value. For users not familiar with this parameter (described in Chapter P01) the recommended value is 0.

Unless the routine detects an error (see next section), IFAIL contains 0 on exit.

6. Error Indicators and Warnings

Errors detected by the routine:–

IFAIL = 1

On entry, K < 2.

IFAIL = 2

On entry, $LX(I) \leq 0$ for some I, $(1 \leq I \leq K)$.

IFAIL = 3

On entry, $L \neq \sum_{I=1}^{K} LX(I)$.

IFAIL = 4

On entry, all the observations were equal.

7. Auxiliary Routines

This routine calls the NAG Library routines G01BCF, G08AEZ, and P01AAF.

8. Timing

The timing is small, and increases with L and K.

9. Storage

There are no internally declared arrays.

10. Accuracy

Basic precision arithmetic is used throughout. For estimates of the accuracy of the significance P, see G01BCF. The χ^2 approximation is acceptable unless k = 3 and l_1, l_2 or $l_3 \leq 5$.

11. Further Comments

If K = 2, the Median test (see G08ACF) or the Mann–Whitney U test (see G08ADF) is more appropriate.

12. Keywords

Kruskal–Wallis One–Way Analysis of Variance, Nonparametric Statistical Test, Distribution – Free Statistical Test.

---

13. Example

This example is taken from page 20 of 'Standard Statistical Calculations' by P.G. Moore, E.A. Shirley, and D.E. Edwards (Pitman, 1972). There are 5 groups of sizes 5, 8, 6, 8, and 8. The data represent the weight gain, in pounds, of pigs from five different litters under the same conditions. The same data is used for the parametric one-way analysis of variance example program for G04AEF.

WARNING: This **single precision** example program may require amendment for certain implementations. The results produced may not be the same. If in doubt, please seek further advice (see **Essential Introduction** to the Library Manual).

13.1. Program Text

```
C     G08AFF EXAMPLE PROGRAM TEXT
C     MARK 8 RELEASE. NAG COPYRIGHT 1979.
C     .. LOCAL SCALARS ..
      REAL H, P
      INTEGER I, IDF, IFAIL, II, K, LX, NHI, NI, NIN, NLO, NOUT
C     .. LOCAL ARRAYS ..
      REAL TITLE(7), W1(35), X(35)
      INTEGER L(5)
```

```
C      .. SUBROUTINE REFERENCES ..
C      G08AFF
C      ..
       DATA NIN, NOUT /5,6/
       READ (NIN,99999) TITLE
       WRITE (NOUT,99997) (TITLE(I),I=1,6)
       WRITE (NOUT,99996)
       READ (NIN,99998) X, L
       LX = 35
       K = 5
       IFAIL = 1
       NLO = 1
       DO 20 I=1,K
          NI = L(I)
          NHI = NLO + NI - 1
          WRITE (NOUT,99995) I, (X(II),II=NLO,NHI)
          NLO = NLO + NI
   20 CONTINUE
       CALL G08AFF(X, LX, L, K, W1, H, P, IFAIL)
       IDF = K - 1
       WRITE (NOUT,99994) H, IDF, P
       IF (IFAIL.NE.0) WRITE (NOUT,99993) IFAIL
       STOP
99999 FORMAT (6A4, A3)
99998 FORMAT (10F3.0/10F3.0/10F3.0/5F3.0/5I2)
99997 FORMAT (4(1X/), 1H , 5A4, A3, 7HRESULTS/1X)
99996 FORMAT (20H KRUSKAL-WALLIS TEST//12H DATA VALUES//9H  GROUP  ,
     * 14H   OBSERVATIONS)
99995 FORMAT (5X, I1, 6X, 10(F3.0, 1X))
99994 FORMAT (//22H TEST STATISTIC         , F8.2/16H DEGREES OF FREE,
     * 6HDOM   , 5X, I3/22H SIGNIFICANCE          , F8.3//)
99993 FORMAT (23H G08AFF FAILS. IFAIL = , I2)
       END
```

13.2. Program Data

```
G08AFF EXAMPLE PROGRAM DATA
 23 27 26 19 30 29 25 33 36 32
 28 30 31 38 31 28 35 33 36 30
 27 28 22 33 34 34 32 31 33 31
 28 30 24 29 30
 5 8 6 8 8
```

13.3. Program Results

```
 G08AFF EXAMPLE PROGRAM RESULTS

 KRUSKAL-WALLIS TEST

 DATA VALUES

  GROUP     OBSERVATIONS
     1        23. 27. 26. 19. 30.
     2        29. 25. 33. 36. 32. 28. 30. 31.
     3        38. 31. 28. 35. 33. 36.
     4        30. 27. 28. 22. 33. 34. 34. 32.
     5        31. 33. 31. 28. 30. 24. 29. 30.
```

```
TEST STATISTIC          10.54
DEGREES OF FREEDOM          4
SIGNIFICANCE            0.032
```

G08AFF – NAG FORTRAN Library Routine Document

NOTE: before using this routine, please read the appropriate implementation document to check the interpretation of bold *italicised* terms and other implementation-dependent details. The routine name may be precision-dependent.

1. Purpose

G08AFF performs the Kruskal–Wallis one-way analysis of variance by ranks on k independent samples of possibly unequal sizes.

2. Specification

```
      SUBROUTINE G08AFF (X,L,LX,K,W,H,P,IFAIL)
C     INTEGER L,LX(K),K,IFAIL
C     real X(L),W(L),H,P
```

---

3. Description

The Kruskal–Wallis test investigates the differences between scores from k independent samples of unequal sizes, the i(th) sample containing l_i observations. The null hypothesis H_0 is that the samples come from the same population, and this is to be tested against the alternative hypothesis H_1 that they come from different populations.

The test proceeds as follows:

1. The pooled sample of all the observations is ranked. Average ranks are assigned to tied scores.
2. The ranks of the observations in each sample are summed, to give the rank sums $R_i, i = 1,...,k$.
3. Kruskal–Wallis' test statistic H is computed as:

$$H = \frac{12}{L(L+1)} \sum_{i=1}^{k} \frac{R_i^2}{l_i} - 3(L+1),$$

where

$$L = \sum_{i=1}^{k} l_i,$$

i.e. the total number of observations. If there are tied scores, H is corrected by dividing by:

$$1 - \frac{\Sigma(t^3-t)}{L^3-L}$$

where t is the number of tied scores in a group and the summation is over all tied groups.

G08AFF returns the value of H, and also an approximation, p, to the probability of such a value being observed, given that H_0 is true. (H approximately follows a χ^2_{k-1} distribution). H_0 is rejected at the 100α% level of significance if $p < \alpha$. The approximation p is acceptable unless $k = 3$ and l_1, l_2 or $l_3 \le 5$ in which case table δ of [1] should be consulted) or $k = 2$ (in which case the median test (see G08ACF) or the Mann–Whitney U test (see G08ADF) is more appropriate).

4. References

[1] SIEGEL, S.
Nonparametric Statistics for the Behavioural Sciences, Chapter 8.
McGraw Hill, 1956.

5. Parameters

X – ***real*** array of DIMENSION at least (L).

Before entry, the elements of X must contain the observations in the K groups. The first l_1 elements must contain the scores in the first group, the next l_2 those in the second group and so on.

Unchanged on exit.

L – INTEGER.

On entry, L specifies the total number of observations.

Unchanged on exit.

LX – INTEGER array of DIMENSION at least (K).

On entry, LX(i) specifies the number of observations l_i in sample i, for $i = 1,...,K$.

$$L = \sum_{i=1}^{k} LX(i).$$

Unchanged on exit.

K – INTEGER.

On entry, K specifies the number of samples, k. K ≥ 2.

Unchanged on exit.

W – ***real*** array of DIMENSION at least (L).

Used as workspace.

H – ***real***.

On exit, H contains the value of the Kruskal–Wallis test statistic.

P – ***real***.

On exit, P contains the approximate significance, p, of the Kruskal–Wallis test statistic.

IFAIL – INTEGER.

Before entry, IFAIL must be assigned a value. For users not familiar with this parameter (described in Chapter P01) the recommended value is 0.

Unless the routine detects an error (see next section), IFAIL contains 0 on exit.

6. Error Indicators and Warnings

Errors detected by the routine:-

IFAIL = 1

On entry, k < 2.

IFAIL = 2

On entry, LX(I) ≤ 0 for some I, (1 ≤ I ≤ k).

IFAIL = 3

On entry, $L \ne \sum_{I=1}^{k} LX(I)$.

IFAIL = 4

On entry, all the observations were equal.

7. Auxiliary Routines

This routine calls the NAG Library routines G01BCF, G08AEZ, and P01AAF.

8. Timing

The timing is small, and increases with nk.

9. Storage

There are no internally declared arrays.

10. Accuracy

Basic precision arithmetic is used throughout. For estimates of the accuracy of the significance P, see G01BCF. The χ^2 approximation is acceptable unless $k = 3$ and l_1, l_2 or $l_3 \le 5$.

11. Further Comments

If K = 2, the Median test (see G08ACF) or the Mann–Whitney U test (see G08ADF) is more appropriate.

12. Keywords

Kruskal–Wallis One–Way Analysis of Variance, Nonparametric Statistical Test, Distribution – Free Statistical Test.

---

13. Example

This example is taken from page 20 of 'Standard Statistical Calculations' by P.G. Moore, E.A. Shirley, and D.E. Edwards (Pitman, 1972). There are 5 groups of sizes 5, 8, 6, 8, and 8. The data represent the weight gain, in pounds, of pigs from five different litters under the same conditions. The same data is used for the parametric one-way analysis of variance example program for G04AEF.

WARNING: This **single precision** example program may require amendment for certain implementations. The results produced may not be the same. If in doubt, please seek further advice (see **Essential Introduction** to the Library Manual).

13.1. Program Text

```
C     G08AFF EXAMPLE PROGRAM TEXT
C     MARK 8 RELEASE. NAG COPYRIGHT 1979.
C     .. LOCAL SCALARS ..
      REAL H, P
      INTEGER I, IDF, IFAIL, II, K, LX, NHI, NI, NIN, NLO, NOUT
C     .. LOCAL ARRAYS ..
      REAL TITLE(7), W1(35), X(35)
      INTEGER L(5)
C     .. SUBROUTINE REFERENCES ..
C     G08AFF
C     ..
      DATA NIN, NOUT /5,6/
```

```
      READ (NIN,99999) TITLE
      WRITE (NOUT,99997) (TITLE(I),I=1,6)
      WRITE (NOUT,99996)
      READ (NIN,99998) X, L
      LX = 35
      K = 5
      IFAIL = 1
      NLO = 1
      DO 20 I=1,K
         NI = L(I)
         NHI = NLO + NI - 1
         WRITE (NOUT,99995) I, (X(II),II=NLO,NHI)
         NLO = NLO + NI
   20 CONTINUE
      CALL G08AFF(X, LX, L, K, W1, H, P, IFAIL)
      IDF = K - 1
      WRITE (NOUT,99994) H, IDF, P
      IF (IFAIL.NE.0) WRITE (NOUT,99993) IFAIL
      STOP
99999 FORMAT (6A4, A3)
99998 FORMAT (10F3.0/10F3.0/10F3.0/5F3.0/5I2)
99997 FORMAT (4(1X/), 1H , 5A4, A3, 7HRESULTS/1X)
99996 FORMAT (20H KRUSKAL-WALLIS TEST//12H DATA VALUES//9H  GROUP ,
     * 14H  OBSERVATIONS)
99995 FORMAT (5X, I1, 6X, 10(F3.0, 1X))
99994 FORMAT (//22H TEST STATISTIC       , F8.2/16H DEGREES OF FREE,
     * 6HDOM   , 5X, I3/22H SIGNIFICANCE          , F8.3//)
99993 FORMAT (23H G08AFF FAILS. IFAIL = , I2)
      END
```

13.2. Program Data

```
G08AFF EXAMPLE PROGRAM DATA
 23 27 26 19 30 29 25 33 36 32
 28 30 31 38 31 28 35 33 36 30
 27 28 22 33 34 34 32 31 33 31
 28 30 24 29 30
 5 8 6 8 8
```

13.3. Program Results

```
 G08AFF EXAMPLE PROGRAM RESULTS

 KRUSKAL-WALLIS TEST

 DATA VALUES

  GROUP   OBSERVATIONS
     1       23. 27. 26. 19. 30.
     2       29. 25. 33. 36. 32. 28. 30. 31.
     3       38. 31. 28. 35. 33. 36.
     4       30. 27. 28. 22. 33. 34. 34. 32.
     5       31. 33. 31. 28. 30. 24. 29. 30.


 TEST STATISTIC          10.54
 DEGREES OF FREEDOM
 SIGNIFICANCE            0.032
```

APPENDIX A2. Photoreduced example of extracts from a Chapter Introduction Document

E04 - Minimizing or Maximizing a Function

1. Scope of the Chapter

An optimization problem involves minimizing a function (called the objective function) of several variables, possibly subject to restrictions on the values of the variables defined by a set of constraint functions. The routines in the NAG Library are concerned with function minimization only, since the problem of maximizing a given function can be transformed into a minimization problem simply by multiplying the function by -1.

It is beyond the scope of this introduction to give other than a brief description of optimization problems and the algorithms used to solve them; however, we have endeavoured to include those aspects of optimization that introduce the reader to the relevant terminology. The purpose of this introduction is threefold: to aid the user in the formulation of the problem, to facilitate algorithm selection and to enable a correct interpretation of the computed results to be made.

2. Background to the Problems

2.1. Types of Optimization Problem

Solution of optimization problems by a single, all-purpose, method is cumbersome and inefficient. Optimization problems are therefore classified into particular categories, where each category is defined by the properties of the objective and constraint functions, as illustrated below.

Properties of Objective Function	Properties of Constraints
Nonlinear	Nonlinear
Sums of squares of nonlinear functions	Sparse linear
Quadratic	Linear
Sums of squares of linear functions	Bounds
Linear	None

For example, a specific problem category involves the minimization of a nonlinear objective function subject to bounds on the variables. In the following sections we define the particular categories of problems that can be solved by routines contained in this Chapter. Not every category is given special treatment in the current version of the library; however, the long-term objective is to provide a comprehensive set of routines to solve problems in all such categories.

INTRODUCTION - E04

2.2. Geometric Representation; Terminology

To illustrate the nature of optimization problems it is useful to consider the following example in two dimensions

$$F(x) = e^{x_1}(4x_1^2 + 2x_2^2 + 4x_1x_2 + 2x_2 + 1) \quad .$$

(This function is used as the example function in the documentation for the unconstrained routines.)

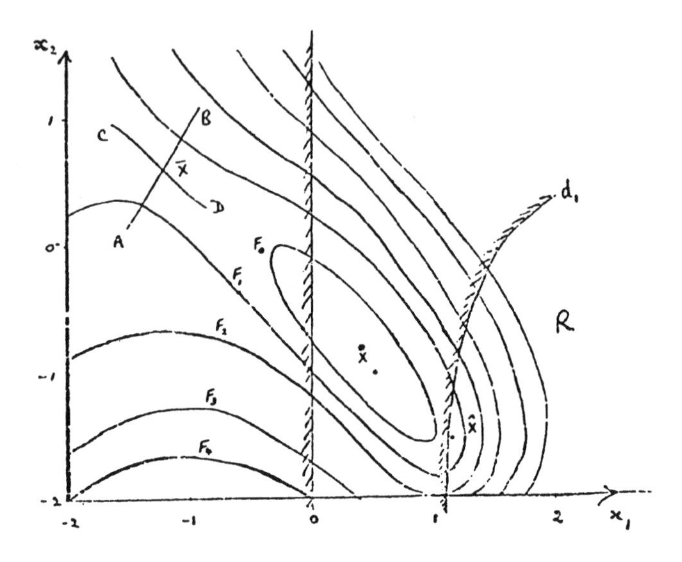

Figure 1

Figure 1 is a contour diagram of F(x). The contours labelled F_0, F_1,.... are isovalue contours, or lines along which the function F(x) takes specific constant values. The point $\overset{*}{x}$ is a local unconstrained minimum, that is, it has a lower function value than all the points adjacent to it. A function may have several such minima. The lowest of the local minima is termed a global minimum. In the problem illustrated in Figure 1, $\overset{*}{x}$ is the only local minimum. The point $\bar{x}$ is said to be a saddle point because it is a minimum along the line AB, but a maximum along CD.

3.3.2. Selection Chart for Unconstrained Problems

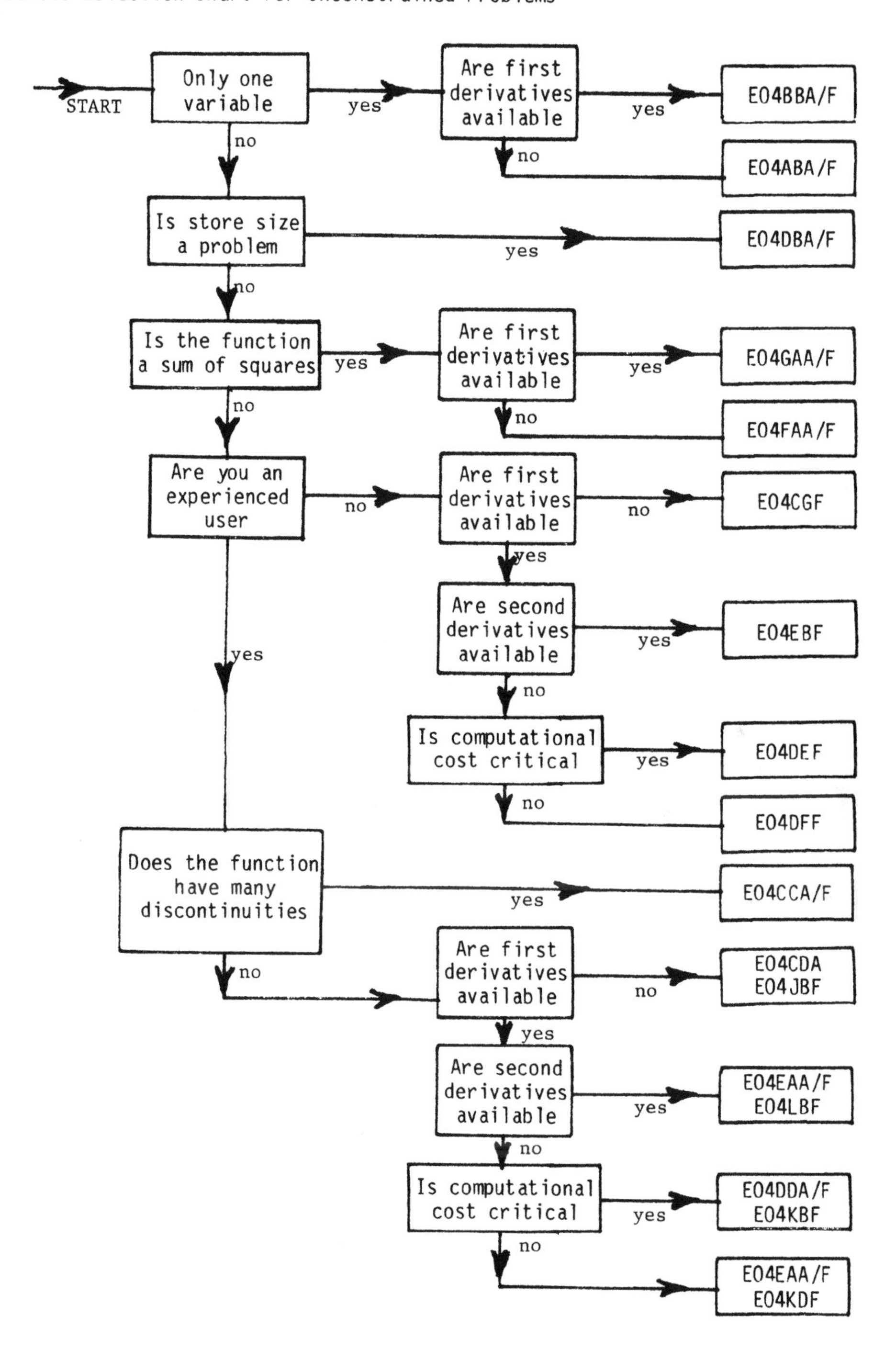

Appendix B – NAG Library (Algol 60 and FORTRAN) Implementations

	NAG Imp. Code	Computer System	Language	Comments
	J1 A	BURROUGHS 6700	ALGOL 60	
	J1 F	BURROUGHS 6700	FORTRAN S.P.	
	U F	CDC 3000L	FORTRAN S.P.	NOT BEYOND MARK 7?
	C2 A4	CDC 6000/LOWER CYBER	ALGOL 60	ALGOL 4 COMPILER
	C2 F	CDC 6000/LOWER CYBER	FORTRAN S.P.	
	C1 A4	CDC 7600	ALGOL 60	SCM AND LCM (ALGOL 4)
	C1 F	CDC 7600	FORTRAN S.P.	SCM AND LCM
	Z F	CRAY-1	FORTRAN S.P.	
	R F	DEC PDP 11	FORTRAN D.P.	IV PLUS COMPILER
	DA A	DEC SYSTEM 10 (KA)	ALGOL 60	
	DA F	DEC SYSTEM 10 (KA)	FORTRAN S.P.	
	DI A	DEC SYSTEM 10 (KI)	ALGOL 60	
	DI F	DEC SYSTEM 10 (KI)	FORTRAN S.P.	
	DL A	DEC SYSTEM 10 (KL)	ALGOL 60	
	DL F	DEC SYSTEM 10 (KL)	FORTRAN S.P.	
	D2 F	DEC SYSTEM 20	FORTRAN S.P.	
	DV F	DEC VAX11	FORTRAN D.P.	
	X F	GEC 4000	FORTRAN D.P.	
	W D	HARRIS/VULCAN	FORTRAN D.P.	
	W F	HARRIS/VULCAN	FORTRAN S.P.	
	HP3 F	HEWLETT PACKARD 3000	FORTRAN D.P.	
†	N1 A	HONEYWELL GCOS	ALGOL 60	
	N1 F	HONEYWELL GCOS	FORTRAN D.P.	
	N2 F	HONEYWELL MULTICS	FORTRAN D.P.	
	B A	IBM 360/370	ALGOL 60	D.P./DELFT COMPILER
	B1 FG	IBM 360/370 (H.E.P.)	FORTRAN D.P.	G COMPILER
	B1 F+	IBM 360/370 (H.E.P.)	FORTRAN D.P.	H EXTENDED COMPILER
	B2 FG	IBM 360/370 (S.E.P.)	FORTRAN D.P.	G COMPILER
	B2 F+	IBM 360/370 (S.E.P.)	FORTRAN D.P.	H EXTENDED COMPILER
	B EG	IBM 360/370	FORTRAN S.P.	G COMPILER
	B E+	IBM 360/370	FORTRAN S.P.	H EXTENDED COMPILER
	H A	ICL 1900*	ALGOL 60	NON-1906A/S
	H F	ICL 1900*	FORTRAN S.P.	NON-1906A/S
	A A	ICL 1906A/S	ALGOL 60	
	A F	ICL 1906A/S	FORTRAN S.P.	
	Q1 F1	ICL 2900 (K)	FORTRAN D.P.	VIA ICL, VME/K
	Q2 AE	ICL 2900 (B)	ALGOL 60	VIA ICL, VME/B/A(EDIN)
	Q2 F	ICL 2900 (B)	FORTRAN D.P.	VIA ICL, VME/B/F OPT
	E F	ICL 4100	FORTRAN S.P.	NOT BEYOND MARK 4?
	F F	ICL SYSTEM 4	FORTRAN D.P.	
	V F	NORD 10/100	FORTRAN S.P.	
	PE F	PERKIN ELMER-32	FORTRAN D.P.	
	P AE	PHILIPS 14/1800	ALGOL 60	
	P F	PHILIPS 14/1800	FORTRAN D.P.	
	MV F	PRIME V MODE	FORTRAN D.P.	VIA PRIME (EUROPE)
	Y F	RANK XEROX 530	FORTRAN D.P.	NOT BEYOND MARK 5?
	S F	SIEMENS BS2000	FORTRAN D.P.	
	T A	TELEFUNKEN TR440	ALGOL 60	
	T F	TELEFUNKEN TR440	FORTRAN S.P.	
	K A	UNIVAC 1100	ALGOL 60 D.P.	NU ALGOL, NOT BEYOND MARK 5?
	K C	UNIVAC 1100	ALGOL 60 S.P.	NU ALGOL, NOT BEYOND MARK 5?
	K F	UNIVAC 1100	FORTRAN D.P.	FTN/FOR(E3)
	K E	UNIVAC 1100	FORTRAN S.P.	FTN/FOR(E3)
†	YS D	XEROX SIGMA	FORTRAN D.P.	
†	YS F	XEROX SIGMA	FORTRAN S.P.	

S.P. = SINGLE PRECISION
D.P. = DOUBLE PRECISION
H.E.P. = WITH HARDWARE EXTENDED PRECISION
S.E.P. = WITH SOFTWARE EXTENDED PRECISION
† = IMPLEMENTATION NOT COMPLETE, JULY 1980.

APPENDIX C.1 Prototype Document in Input Form

```
*COY
*DIY
*UPTODATE G08AFF
*DA "DECEMBER 1979"
*CA "G08 - Nonparametric Statistics" "G08AFF" "1744/0"
*WA *H1
*NAG G08AFF performs the Kruskal-Wallis one-way analysis of
variance by ranks on k independent samples of possibly unequal
sizes. *N
*H2
*C _____ SUBROUTINE
*C *NAG              G08AFF (X, L, LX, K, W, H, P, IFAIL)          *N
*C C____ INTEGER L, LX(K), K, IFAIL                                 *N
*C C____ *SP real
*C           X(L), W(L), H, P                                      *N
*SK _____ SUBROUTINE
*SK *NAG             G08AFF (X, L, LX, K, W, H, P, IFAIL)          *N
*SK C____ INTEGER L, LX(K), K, IFAIL                                *N
*SK C____ *SP real
*SK          X(L), W(L), H, P                                      *N
*H3
The Kruskal-Wallis test investigates the differences between
scores from [k] independent samples of unequal sizes, the
[i@F3(th)] sample containing [l SUB [i]] observations.  The
hypothesis under test, [H SUB 0], often called
the null hypothesis, is that the samples come from the same
population, and this is to be tested against the alternative
hypothesis [H SUB [1]] that they come from different populations.
*N *IL5 *PL3 *HL The test proceeds as follows: *N *PL3 *HL
1._The pooled sample of all the observations is ranked.
Average ranks are assigned to tied scores. *N *HL *PL3
2._The ranks of the observations in each sample are summed, to
give the rank sums [R SUB [i],_i_=_1,...,k.] *N *PL3 *HL
3._Kruskal-Wallis' test statistic H is computed as: *N *HL
[@F3H _=_12 OVER [@F3L (@F3L+1)] _SIGMA FROM i=1 TO k
[R SUB i SUP 2] OVER [l SUB i]__-_
3(@F3L +1),] *QC *IL5 *HL
where *N *HL
[@F3L _=_ SIGMA FROM i=1 TO k [l SUB i] ,] *QC *IL5 *HL
i.e. the total number of observations.  If there are tied scores,
H is corrected by dividing by: *N *HL
[1_-_[SIGMA (t SUP 3 - t)] OVER [@F3L SUP 3 - @F3L]] *QC *IL4 *HL
where [t] is the number of tied scores in a group and the summation
is over all tied groups. *N *IL2 *HL
*NAG G08AFF returns the value of H, and also an approximation,
[p], to the probability of a value of at least H being observed,
[H SUB [0]] is true. (H approximately follows a
[chi SUB [k-1] SUP 2] distribution).  [H SUB [0]] is rejected
by a test of chosen size [@GA] if [p_@AI_@GA.]
The approximation [p] is acceptable unless [k_=_3] and
[l SUB [1] ,l SUB [2]] or [_l SUB [3] _@AM_5] (in which case table
0 of "[1]" should be consulted) or [k_=_2] (in which case
the Median test (see *NAG G08ACF ) or the Mann-Whitney U test
(see *NAG G08ADF ) is more appropriate). *N *IL0
```

Appendix C.1
page 2

```
*H4
*IL4 *PL4 "[1]"_SIEGEL, S. *N
Nonparametric Statistics for the Behavioral Sciences, Chapter 8. *N
McGraw Hill, 1956. *N
*H5
*% X - *SP real array of DIMENSION at least (L). *N
Before entry, the elements of X must contain the observations in the
K groups.  The first [l SUB [1]] \ l(1) \ elements must contain
the scores in the first group, the next \ l(2) \ [l SUB 2]
those in the second group, and so on.  *N *HL Unchanged on
exit. *N
*% L - INTEGER. *N
On entry, L must specify the total number of observations. *N
*HL Unchanged on exit. *N
*% LX - INTEGER array of DIMENSION at least (K). *N
Before entry, [@F3LX (i)] \ LX(i) \ must contain the number of
observations [l SUB i] \ l(i) \
in sample [i] \ i \ , for [i] \ i \ = 1,...,K. *N *HL
*SK L [ = [@CA FROM i=1 TO k _ ROMAN LX (i)]].
*C L_=_SUM from i_=1 to K of LX(i).
*QC *IL2 *HL
 Unchanged on exit. *N
*% K - INTEGER. *N
On entry, K must specify the number of samples [,_k]. *N K *GE 2. *N
*HL Unchanged on exit. *N
*% W - *SP real array of DIMENSION at least (L). *N
Used as workspace. *N
*% H - *SP real . *N
On exit, H contains the value of the Kruskal-Wallis test statistic.
*N *% P - *SP real . *N
On exit, P contains the approximate significance [,_p,] of the
Kruskal-Wallis test statistic. *N
*IFAIL *H6
*% Errors detected by the routine:- *N *LS
*% IFAIL = 1 *N On entry, K *LT 2. *N
*% IFAIL = 2 *N On entry, LX(I) *LE 0 for some I, (1 *LE I *LE K).
*N *% IFAIL = 3 *N On entry, L
[@A7 SIGMA FROM @F3I=1 TO @F3K _ ROMAN LX(I) .]
\ *NE SUM from I=1 to K of LX(I). \
*N *% IFAIL = 4 *N On entry, all the observations were equal.
*N *H7
This routine calls the NAG Library routines *NAG G01BCF ,
G08AEZ, and *NAG P01AAF . *N
*H8
The timing is small, and increases with L and K. *N
*H9
There are no internally declared arrays. *N
*H10
*SP Basic_precision arithmetic is used throughout.  For estimates
of the accuracy of the significance P, see *NAG G01BCF .  The
[chi SUP 2] approximation is acceptable unless [k_=_3 ] and
[l SUB [1] ,l SUB [2]] or [l SUB [3] _@AM_5]. *N
*H11
If K = 2, the Median test (see *NAG G08ACF ) or the Mann-Whitney
U test (see *NAG G08ADF ) is more appropriate. *N
*H12
```

Appendix C.1
page 3

```
Kruskal-Wallis One-Way Analysis of Variance,
Nonparametric Statistical Test,
Distribution - Free Statistical Test. *N
*H13
This example is taken from page 20 of *LQ Standard Statistical
Calculations *TQ by P.G. Moore, E.A. Shirley, and D.E. Edwards
(Pitman, 1972).  There are 5 groups of sizes 5, 8, 6, 8, and 8.  The
data represent the weight gain, in pounds, of pigs from five
different litters under the same conditions.  The same data
is used for the parametric one-way analysis of variance example
program for *NAG G04AEF . *N
*W13 *H13.1
"C      G08AFF EXAMPLE PROGRAM TEXT" *N
"C      MARK 8 RELEASE. NAG COPYRIGHT 1979." *N
"C      .. LOCAL SCALARS .." *N
"       REAL H, P" *N
"       INTEGER I, IDF, IFAIL, II, K, LX, NHI, NI, NIN, NLO, NOUT" *N
"C      .. LOCAL ARRAYS .." *N
"       REAL TITLE(7), W1(35), X(35)" *N
"       INTEGER L(5)" *N
"C      .. SUBROUTINE REFERENCES .." *N
"C      G08AFF" *N
"C      .." *N
"       DATA NIN, NOUT /5,6/" *N
"       READ (NIN,99999) TITLE" *N
"       WRITE (NOUT,99997) (TITLE(I),I=1,6)" *N
"       WRITE (NOUT,99996)" *N
"       READ (NIN,99998) X, L" *N
"       LX = 35" *N
"       K = 5" *N
"       IFAIL = 1" *N
"       NLO = 1" *N
"       DO 20 I=1,K" *N
"          NI = L(I)" *N
"          NHI = NLO + NI - 1" *N
"          WRITE (NOUT,99995) I, (X(II),II=NLO,NHI)" *N
"          NLO = NLO + NI" *N
"    20 CONTINUE" *N
"       CALL G08AFF(X, LX, L, K, W1, H, P, IFAIL)" *N
"       IDF = K - 1" *N
"       WRITE (NOUT,99994) H, IDF, P" *N
"       IF (IFAIL.NE.0) WRITE (NOUT,99993) IFAIL" *N
"       STOP" *N
"99999 FORMAT (6A4, A3)" *N
"99998 FORMAT (10F3.0/10F3.0/10F3.0/5F3.0/5I2)" *N
"99997 FORMAT (4(1X/), 1H , 5A4, A3, 7HRESULTS/1X)" *N
"99996 FORMAT (20H KRUSKAL-WALLIS TEST//12H DATA VALUES//9H  GROUP "
"," *N
"     * 14H  OBSERVATIONS)" *N
"99995 FORMAT (5X, I1, 6X, 10(F3.0, 1X))" *N
"99994 FORMAT (//22H TEST STATISTIC        , F8.2/16H DEGREES OF"
"FREE," *N
"     * 6HDOM   , 5X, I3/22H SIGNIFICANCE          , F8.3//)" *N
"99993 FORMAT (23H G08AFF FAILS. IFAIL = , I2)" *N
"       END" *N
*H13.2
```

Appendix C.1
page 4

```
"G08AFF EXAMPLE PROGRAM DATA" *N
" 23 27 26 19 30 29 25 33 36 32" *N
" 28 30 31 38 31 28 35 33 36 30" *N
" 27 28 22 33 34 34 32 31 33 31" *N
" 28 30 24 29 30" *N
" 5 8 6 8 8" *N
*H13.3S
" " *N
" " *N
" " *N
" " *N
" G08AFF EXAMPLE PROGRAM RESULTS" *N
" " *N
" KRUSKAL-WALLIS TEST" *N
" " *N
" DATA VALUES" *N
" " *N
"  GROUP    OBSERVATIONS" *N
"     1      23. 27. 26. 19. 30." *N
"     2      29. 25. 33. 36. 32. 28. 30. 31." *N
"     3      38. 31. 28. 35. 33. 36." *N
"     4      30. 27. 28. 22. 33. 34. 34. 32." *N
"     5      31. 33. 31. 28. 30. 24. 29. 30." *N
" " *N
" " *N
" TEST STATISTIC           10.54" *N
" DEGREES OF FREEDOM           4" *N
" SIGNIFICANCE             0.032" *N
" " *N
" " *N
" " *N
**END OF G08AFF
$
```

APPENDIX C.2 Prototype Document after Extraction

```
*UPTODATE G08AFF
G08AFF - NAG FORTRAN ROUTINE SUMMARY
======


IMPORTANT: For a complete specification of the use of this
routine see the NAG FORTRAN  Library  Manual. Terms marked
// ... // may be implementation dependent.


A. Purpose
==========

 //G08AFF//  performs the Kruskal-Wallis  one-way analysis
 of variance by ranks on k independent samples of possibly
 unequal sizes.

B. Specification
================

       SUBROUTINE //G08AFF// (X, L, LX, K, W, H, P, IFAIL)
 C     INTEGER L, LX(K), K, IFAIL
 C     //real// X(L), W(L), H, P

C. Parameters
=============


X - //real// array of DIMENSION at least (L).

  Before  entry,  the  elements  of  X  must  contain  the
  observations  in the K groups.  The first l(1)  elements
  must  contain  the scores in the first  group,  the next
  l(2) those in the second group, and so on.

  Unchanged on exit.

L - INTEGER.

  On   entry,   L   must   specify  the  total  number  of
  observations.

  Unchanged on exit.

LX - INTEGER array of DIMENSION at least (K).

  Before    entry,   LX(i)  must  contain  the  number  of
  observations l(i) in sample i , for i = 1,...,K.

              L = SUM from i =1 to K of LX(i).

  Unchanged on exit.
```

Appendix C.2 (cont'd)

K - INTEGER.

On entry, K must specify the number of samples K.GE.2.

Unchanged on exit.

W - //real// array of DIMENSION at least (L).

Used as workspace.

H - //real// .

On exit, H contains the value of the Kruskal-Wallis test statistic.

P - //real// .

On exit. P contains the approximate significance of the Kruskal-Wallis test statistic.

IFAIL - INTEGER.

Before entry, IFAIL must be assigned a value. For users not familiar with this parameter (described in Chapter P01) the recommended value is 0.

Unless the routine detects an error (see next section), IFAIL contains 0 on exit.

D. Error Indicators and Warnings
================================

Errors detected by the routine:-

IFAIL = 1

On entry, K.LT.2.

IFAIL = 2

On entry, LX(I).LE.0 for some I, (1.LE.I.LE.K).

IFAIL = 3

On entry, L.NE.SUM from I=1 to K of LX(I).

IFAIL = 4

On entry, all the observations were equal.

END OF G08AFF FORTRAN SUMMARY - MARK 8: DECEMBER 1979

** END OF G08AFF

APPENDIX D

Appendix D. An example of mathematics typesetting

This example is taken from the book 'Computer Peripherals and Typesetting' by A.H.Phillips and which is published by H.M.S.O. It has been reset here using TSSD and Harwell's Photon mark I Pacesetter. It shows that TSSD can set mathematics comparable with the best in the printing industry although perhaps not quite as good as the very best hand work. There are one or two places where manual intervention would have improved the page although using TSSD was considerably cheaper than if it were done by hand and also reformatting and making corrections become almost trivial operations.

$$
\begin{aligned}
G_j(y) &= -\sum_{\nu=0}^{2s-1} \frac{B_\nu C_{4s-\nu}^{\nu} y^{4s-2\nu}(p^{4s-\nu-1}+q^{4s-\nu-1})}{(4s-\nu-1)(4s-\nu)} \\
&\quad - \frac{B_{2s}}{2s(2s-1)}[p^{2s-1}+q^{2s-1}-(pq)^{2s-1}], \quad j=4s-2,
\end{aligned} \tag{29}
$$

and

$$
T_j(y) = \sum \frac{G_1^{\nu_1}(y)\ldots G_j^{\nu_j}(y)}{\nu_1!\ldots\nu_j!} \tag{30}
$$

is a polynomial of degree $3j$ which is even when j is an even integer.

Theorem 6. *The asymptotic expansion*

$$
\sum_{i=\mu_0+1}^{\mu_0+m} C_n^i p^i q^{n-i} = \Phi(y_2)-\Phi(y_1)+\sum_{j=1}^{\mu-1}\frac{Q_j}{(\sqrt{npq})^j}+O\left(\frac{1}{n^{\mu/2}}\right) \tag{31}
$$

holds as $n\to\infty$ *for any fixed* θ *uniformly for* $a\leqslant y_1\leqslant y_2\leqslant b$, *where*

$$
y_1 = \frac{\mu_0+\theta-np}{\sqrt{npq}}, \qquad y_2 = \frac{\mu_0+m+\theta-np}{\sqrt{npq}},
$$

$$
\Phi(y) = \int_{-\infty}^{y}\varphi(y)\,dy, \qquad \varphi(y) = \frac{e^{-y^2/2}}{\sqrt{2\pi}},
$$

$$
\begin{aligned}
Q_j &= -\frac{B_j(\theta)}{j!}H_{j-1}(y)\varphi(y)\Big|_{y_1}^{y_2} + \int_{y_1}^{y_2}\varphi(y)T_j(y)\,dy \\
&\quad + \sum_{\nu=1}^{j-1}\frac{(-1)^\nu B_\nu(\theta)}{\nu!}\frac{d^{\nu-1}}{dy^{\nu-1}}\varphi(y)T_{j-\nu}(y)\Big|_{y_1}^{y_2},
\end{aligned}
$$

and

$$
H_j(y) = (-1)^j e^{y^2/2}\frac{d^j}{dy^j}e^{-y^2/2}
$$

is the j-th Hermite polynomial.

PROOF. By (28),

$$
\begin{aligned}
\sum_{i=\mu_0+1}^{\mu_0+m} C_n^i p^i q^{n-i} &= \sum_{i=1}^{m}\frac{1}{\sqrt{2\pi npq}}\exp\left\{-\frac{(\mu_0+i-np)^2}{2npq}\right\} \\
&\quad \times\left\{1+\sum_{j=1}^{\mu-1}\frac{1}{(\sqrt{npq})^j}T_j\left(\frac{\mu_0+i-np}{\sqrt{npq}}\right)\right\} \\
&\quad + m\cdot O\left(\frac{1}{n^{(\mu+1)/2}}\right).
\end{aligned}
$$

Transportable Numerical Software

B. Ford

1. The importance of Numerical Software

It is often held that numerical computation accounts for only a rather small proportion of present-day computing. However this belief is mistaken: a recent study concluded that "numerical computation accounts for about 50% of the computing expenditures in the United States" [1]. Therefore the problems of numerical computation should be of interest not only to those directly concerned with its results but also to anyone with a general interest in computing, and in particular to computer manufacturers and the designers of hardware and software.

Numerical Software is concerned with the manipulation, transformation and computation of floating point numbers. There are three distinct steps in the development of numerical software.

1. the design of the algorithm;
2. its realisation as a source language subprogram;
3. the testing of the compiled code on a given configuration and its detailed documentation for that configuration.

The first stage lies essentially in the provinces of pure or applied mathematics and numerical analysis, which are both ancient and respected disciplines. This stage has been stimulated, and even revolutionized, over the last 30 years by the new practical possibilities which computers have opened up. (It has been estimated that to solve certain 3-dimensional partial differential equations, over the period 1945-75 "the gain in speed from algorithm improvement exceeds the gain from hardware improvement" [2].)

But how are these improvements in algorithm design to be made available to the average user? That is the task of the second and third stages in the development of numerical software which for many years did not receive proper recognition. But the last 10 years have witnessed the following developments: many conferences devoted to this area; a journal published (ACM Transactions on Mathematical Software); an IFIP Working Group established (WG 2.5 on Numerical Software); funding for specific projects (e.g. EISPACK, LINPACK); and the availability of three general purpose subroutine libraries (IMSL, NAG, PORT).

These three libraries have all been implemented on many different ranges of computers. They are not truly portable. It is not possible to compile the same source text without change and achieve acceptable performance on each machine range. But all three library developers have endeavoured to minimize the changes required to adapt the software to a particular computing environment and to implement these changes as far as possible by automatic transformations; this is what we mean by *transportable* software.

The purpose of this talk is to describe the problems to be faced in developing transportable numerical software and the approach adopted by the NAG Project and independently by Bell Laboratories in particular to overcome them.

2. Definitions

The particular characteristics of numerical computing necessitate careful specification of the configuration within which such programs are executed. Each *configuration* is a defined combination of machine hardware, operating system, compiler and compiler libraries.

We have discussed earlier the three steps in the development of numerical software and in particular the distinction between an algorithm and a routine (which is a realisation of the algorithm). The adaptability of an algorithm is the degree to which it can be configured to function efficiently and accurately on various configurations. An *adaptable algorithm* is one which can be configured to compute efficiently and accurately in any chosen computing environment. Waite has used the word *adaptability* in the context of software [20]. We have chosen to constrain the definition to algorithms.

Brown in his pioneering paper on software portability suggested that 'a program or programming system is called portable if the effort required to move it into a new environment is much less than the effort that would be required to program it for the new environment' [17]. We believe that the term *portable* should refer to a quality of a source text rather than a measure of manpower (although obviously the two are related). Any claims about the performance of a given source code should be made in the context of

(a) a clearly defined domain of application, that is, it is portable between configurations A and B,
(b) satisfactory completion of a pre-determined test of accuracy and of efficiency
and (c) a retained ability to solve a given set of numerical problems (unimpaired algorithmic capability).

A program is *portable* over a given range of configurations if without any alteration, it can compile and run to satisfy specified performance criteria on that range.

In transferring a program between members of a given range of configurations, some changes may be necessary to the base version before it satisfies specified performance criteria on each alternative configuration. A program is *transportable* if the modifications to achieve a satisfactory performance on each of configurations A and B are

(a) limited in number (and extent and complexity)
(b) are capable of automatic implementation by a processor
and (c) are such that the original and processed programs are recognisably similar.

The ease with which the processor is written, and the extent to which point (a) is met, reflect the degree of transportability of the program within the specified domain [13].

3. Adaptability of Algorithms

When discussing the portability of numerical software perhaps the first question that should be asked is "will the algorithm realised by the software be adaptable?" This is particularly relevant if we wish our algorithm library, given well-conditioned data, to return a result reflecting the precision of the used configuration.

We have seen that an algorithm is adaptable if it can be expressed in such a way that, relative to some specifications, the algorithm performs in an equivalent manner on a wide range of computing environments. The key phrases in this concept are 'in an equivalent manner' and 'relative to some specification'. We consider here a rather special specification of performance of the algorithm in that this specification must be in terms of the environment in which the algorithm is expected to run. In this context, the phrase 'in an equivalent manner' then means that the performance of the algorithm is measured against this environment-dependent specification. For instance, the usual Gaussian elimination algorithm for factorization of a matrix is an adaptable algorithm, because: 1) the specification of its behaviour, namely determining the factors of a matrix "near" to the original matrix (where "near" is defined in terms of the precision of the arithmetic unit), is expressed in terms of the environment: and 2), it can be proven that at least using Wilkinson's model [14] for the behaviour of rounding errors, this factorization process satisfies this specification. An example of an algorithm that is not adaptable would be an iteration procedure which terminates when the difference of two iterates is less that 10**(-8) in magnitude. For an environment which has nine or more decimal digits in its arithmetic operations, the algorithm may guarantee (as the result of an error analysis) that the last iterate approximates the solution of the problem to within eight digits. For an environment which has fewer than nine digits, the behaviour of the program may be unpredictable for it may not terminate, or may terminate with no accuracy guaranteed.

Adaptable algorithms in numerical software fall into two categories; 1) non-iterative (direct) processes, as exemplified by Gaussian elimination, which in order to meet their specification, operate on numerical data in a fixed manner independent of the computing environment; and 2) iterative (indirect) processes, which in order to meet their specification, vary the manner in which the numerical data is manipulated depending upon the environment. In order to prepare software, based on such algorithms, that can be easily moved about, the dependence of the algorithm on its environment is expressed in terms of machine parameters such as relative machine precision, the base of the machine arithmetic, the largest and smallest positive representable numbers, and several others (cf. [18], [19], [3] and papers referenced by these).

In terms of transportable software, adaptable algorithms take on a new dimension. In the previous discussion, the dependence of the algorithms on the configuration were expressed simply in terms of variables or parameters in the program. In the environment of transportable software, the dependence on the environment can be expressed by use of special constructs recognised by program analyzers. A good example of such an approach is given in Schonfelder [15]. There, a package of programs is developed to evaluate special functions.

Truncated Chebyshev expansions which yield 30 digits of precision are derived in a machine environment which handles such high precision computations. Then programs for other configurations that use fewer digits of precision are generated from this database of 30 digit approximations. The programs so produced are sufficiently accurate and efficient for the applications the author has in mind. The resulting accuracy and efficiency are not as good as could be obtained using other approximations such as minimax rational approximations but are satisfactory for his applications.

Another example of using the transportable approach occurs in the development of LINPACK [16]. Here, versions of linear equation solvers are derived from one common form; that is, from single precision complex versions of the programs, single and double precision versions for real input data as well as double precision versions for complex data are derived.

4. Source language for Subroutines

The second major stage in the development of numerical software is realisation of an algorithm in a source language. The first essential requirement for our software is of course a suitable high-level language in which to code it. The promulgation of the ANSI FORTRAN Standard in 1966 was one cause of the proliferating activity in numerical software over the succeeding years. Most numerical software is now written in FORTRAN and this paper will largely be concerned with FORTRAN, but the usefulness of other languages must not be overlooked. (NAG in fact is actively developing libraries in FORTRAN, Algol 60 and Algol 68.)

However, despite the acceptance of the standard and the widespread availability of "standard-conforming" compilers, there are still many purely linguistic obstacles to the portability of FORTRAN software:

1. Some compilers do not implement the standard completely or correctly, e.g. they impose restrictions on the ordering of specification statements, or they violate the integrity of parenthesized expressions;

2. The standard is not always clear, e.g. is it permissible to call a subroutine with an adjustable array dimension set to zero, provided that the array in question is not referenced in that call of the subroutine?

3. The standard is self-confessedly incomplete, e.g. it does not specify a limit to the number of arguments of a subroutine or the number of nested parentheses in an expression. "Standard-conforming" compilers may, and do, impose restrictions on these aspects of the complexity of a program.

4. The standard is permissive: it permits compilers to add features not described in the standard; or even to implement features prohibited in the standard (the standard merely says that such features have no standard interpretation). Thus various "standard-conforming" compilers allow the initial parameter of a DO-loop to exceed the terminal parameter - and interpret this in different ways.

5. Programmers are frequently unaware whether their programs conform to the standard or not; and their compilers may not be capable of providing such information (this is particularly true of run-time, as opposed to compile-time, features).

To avoid such obstacles we must code our software in a disciplined style [3], using a carefully selected subset of standard FORTRAN which experience has shown as far as possible to be acceptable, and to have the same interpretation, on all computers of interest.

NAG uses the PFORT Verifier [4] to check that all its software conforms to the PFORT subset: this subset meets almost all of NAG's known requirements.

Other software tools used by NAG automatically *impose standards*. DECS uses output from the PFORT verifier to standardize all type and dimension specification statements, while POLISH [5] standardizes the overall layout of each program unit. Such standardization not only improves the appearance, readability and maintainability of the software, but also provides a convenient base for subsequent transformation by other software tools.

One of the most important transformations is the generation of a *double precision* version of the source text, which is required for many machines on which single precision does not provide adequate accuracy. NAG uses a comparatively unsophisticated tool called APT [6] to perform this transformation which is on the whole straightforward to specify and implement provided that the basic single precision version has been suitably standardized, in particular if all REAL variables, arrays and functions are explicitly declared in type specification statements. However further defects in the FORTRAN 66 standard are encountered. Some single precision constructs have no double precision analogues in the standard, especially; F output conversion; the intrinsic functions FLOAT and AINT; any features involving the COMPLEX data type.

5. Problems with Machine Hardware

In numerical computation, even correct and apparently portable FORTRAN statements may have surprising effects. The statement

```
IF (A.NE.0.0) B = 1.0/A
```

may yet cause division by zero.
The logical expression

```
1.25*A .LE. A
```

may have the value .TRUE.

The logical expression

```
1.0/A .GT. B/C
```

may have the value .TRUE. although mathematically (1/A) < (B/C) (CDC 7600 with $A = 1+2^{-46}$, $B = 1-3x2^{-4}$, $C = 1-3x2^{-4}+3x2^{-46}$).

Such pathological examples undermine the rigorous basis on which we would like our software to be founded; or else they imply that to achieve rigour, we must sacrifice some of the range and precision of floating-point numbers which the computer offers.

But much more fundamental problems arise from the fact that in numerical computation on <u>any</u> conventional machine we must work with a system of floating-point numbers of <u>finite</u> precision and <u>finite</u> range, whereas the mathematical method to be implemented may assume that numbers may be of arbitrary size and accuracy. Moreover different machines offer different ranges and precisions of floating-point numbers, use a different base for their representation and many subtly different algorithms in the hardware for the fundamental arithmetic operations (addition, subtraction, multiplication and division). Our goal is that our algorithms should compute on each machine as accurately, efficiently and robustly as the properties of the machine allow. How can this be achieved or approached?

Sometimes problems can be avoided by recasting the source-text with sufficient care. An example due to Cody [7,8] illustrates this well. The straightforward code

```
      Z = SQRT(X**2 + Y**2)                    (1)
```

has the disadvantage that overflow or underflow may occur in evaluating the argument of SQRT even when X,Y and the correct value of Z, are well within the range of floating-point numbers. This may either cause an abnormal termination or lead to incorrect results without warning. These problems can be avoided by rewriting the code thus:

```
      Z = AMAX1(ABS(X),ABS(Y))
      IF (Z.EQ.0.0.) GO TO 10                  (2)
      W = AMIN1(ABS(X),ABS(Y))
      Z = Z * SQRT(1.0 + (W/Z)**2)
   10 CONTINUE
```

But now we have introduced a new potential source of error: on machines which represent floating-point numbers to base 8 or 16, numbers in the range [1.0, 2.0) such as the argument of SQRT - have 2 or 3 bits less accuracy than is achieved in other parts of the range. The loss of accuracy can in turn be avoided by recasting the code yet again and replacing the fourth statement in (2) by:

```
      Z = Z * (SQRT(0.25 + 0.25*(W/Z)**2))  /  0.5)
```

This code gives results as accurate as (1), avoids overflow and returns accurate results if (W/Z)**2 underflows and is set to zero.

Is such extra care - and the extra complexity of the resultant code - justifiable or desirable? That is a matter of judgement for software developers, but in certain contexts it may certainly be justifiable e.g. in the approximation of special functions or the determination of a Givens rotation. In the long term it is not desirable and should not be necessary. Software measures to overcome hardware flaws have been dubbed "algorithmic pollution" by Reinsch [9], who has suggested launching a crusade of "unsafe at any speed" against hardware designers who appear to be more interested in making their machines faster than in giving them soundly based, generally accepted and well-documented arithmetic characteristics. We welcome the efforts currently being made under the auspices of the I.E.E.E. to define a standard for floating-point arithmetic.

6. Parameterisation of the Environment

Many algorithms will be critically influenced by the arithmetic and other characteristics of the configuration on which they are being executed, and must therefore adapt to these characteristics in order to achieve the desired standards of accuracy, efficiency and robustness on that configuration. Certain key parameters of the arithmetic behaviour have in general been found to provide sufficient information.

Table 1. Suggested Parameters for Transportable Numerical Software

Characteristic Name	Definition	Explicit name		
		INTEGER	REAL	DOUBLE PRECISION
ARITHMETIC SET				
radix	Base of the floating-point number system.	—	SRADIX	DRADIX
mantissa length	Number of base-RADIX digits in the mantissa of a stored floating-point number (including, for example, the implicit digit when the first bit of the mantissa of a normalized floating-point number is not stored).	—	SDIGIT	DDIGIT
relative precision	The smallest number x such that $1.0-x < 1.0 < 1.0+x$ where $1.0-x$ and $1.0+x$ are the stored values of the computed results.	—	SRELPR	DRELPR
overflow threshold	The largest integer i such that all integers in the range $[-i,i]$ belong to the system of INTEGER numbers.	IOVFLO	—	—
	The largest number x such that both x and $-x$ belong to the system of REAL (DOUBLE PRECISION) numbers.	—	SOVFLO	DOVFLO
underflow threshold	The smallest positive real number x such that both x and $-x$ are representable as elements of the system of REAL (DOUBLE PRECISION) numbers.	—	SUNFLO	DUNFLO
symmetric range	The largest integer i such that the arithmetic operations □ are exactly performed for all integers a,b satisfying $\|a\|, \|b\| \le i$ provided that the exact mathematical results of a □ b do not exceed i in absolute value.	IRANGE	—	—
	The largest real number x such that the arithmetic operations • are correctly performed for all elements a,b of the system of REAL (DOUBLE PRECISION) numbers, provided that a,b and the exact mathematical result a • b do not have an absolute value outside the range $[1/x,x]$.	—	SRANGE	DRANGE

REPRESENTATION AND ARITHMETIC OPERATIONS

We use the systems of INTEGER, REAL, AND DOUBLE PRECISION numbers as described in the ANSI Fortran standard.[1] For the system of INTEGER numbers we consider the arithmetic operation a □ b, where □ belongs to {+,−,×}, and the monadic operation −a. The computed and stored result can be computed exactly. For the systems of REAL and DOUBLE PRECISION numbers we consider the arithmetic operation a•b, where • belongs to {+,−,×,/}, and the monadic operation −a, to be performed correctly if the computed and stored result can be expressed exactly as a(1+E') • b(1+E") and −a(1+E'"), respectively, where |E'|, |E"|, and |E'"| are at most comparable to relative precision.

The names are those proposed by IFIP WG 2.5 [10].

These parameters have obvious uses in tests for convergence, tests for negligibility, tests to avoid overflow and the damaging consequences of underflow. They may be considered as defining a conceptual machine, for which the programmer writes his source-language routine.

There is still some discussion about how to make the values of the parameters available - whether as the values of configuration-dependent functions (NAG, PORT) or as numerical constants automatically substituted for symbolic names in the source-text (IMSL). But there is general agreement that a few parameters define an adequate, if incomplete working model of the arithmetic behaviour of present-day computers. (If the hardware is "well-behaved," the model may even be complete.)

7. Software Practice and Experience

The general approach of NAG (and independently of Bell Laboratories) to numerical software development is adaptable algorithms coded as transportable software. The transportable software is written in a carefully defined subset FORTRAN and is based upon key parameters of configuration arithmetic behaviour.

The following diagram shows the organisation within the NAG Project for producing a new Mark of the Library:

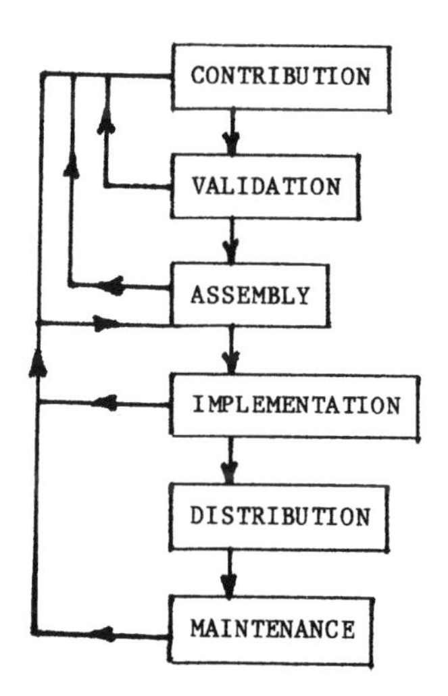

Contributors develop new routines (with test programs and documentation). Validators are asked to certify in particular the selection and design of algorithms. The task of library assembly is performed by the NAG Central Office: our aim is to produce a standardized base version of the library software, which conforms to the required standards and has passed through a uniform testing procedure. Initially we work on a single precision version; the preparation of a double precision base version (and its variants) is a subsequent phase. Implementors then test the Library on each particular machine range, making any changes that are necessary, finally producing compiled and tested code for distribution to user sites.

It might be thought that the assembly of new software into the Library should be a straightforward task. But it is not, and a recent case study [12] has analyzed some of the difficulties. This case study concerned the assembly of a suite of routines and test programs - amounting to 14000 lines of FORTRAN altogether - for the Mark 7 FORTRAN Library. 34 problems (or types of problem) were encountered, which were classified as follows:

a.) Use of non-standard FORTRAN - 8 problems (5 of which involved run-time features or communication between program units).

b.) Undesirable programming style (which could cause difficulties in using or maintaining the software) - 11 problems.

c.) Semantic or pragmatic programming errors (which came to light when the software was tested on different machines) - 5 problems.

d.) Unsatisfactory design of routines, examples or test programs - 5 problems.

e.) Error in the existing NAG Library - 1 problem.

f.) Errors in software tools (compounded by human errors in their application) - 4 problems.

Classified from a different point of view 13 problems came to light during the application of software tools; 6 during visual examination of the software; 15 during compilation and execution on 3 different machines (ICL 1906A, CDC 7600 and IBM 370).

All these problems were of course remedied. They remind us that it is not enough to specify standards and guidelines; we must have adequate (i.e. automatic) means of ensuring that they are adhered to. Increasingly NAG is making software tools such as the PFORT verifier available to contributors, but we cannot expect them to have access to such a wide range of facilities and machine environments as the Central Office.

8. Obstacles to Transportability

Again it might be thought that once the base versions of the Library had been prepared in accordance with our declared approach, a separate implementation phase should not be necessary, or should at most be a straightforward process of assigning values to the machine parameters, compiling the Library and assessing the test results. However NAG has always emphasized the need, in real life, for a corrective phase in implementation [12], because we nearly always encounter unanticipated problems, which might be regarded as "obstacles to transportability".

a) Errors in NAG Library Routines

It is not surprising that testing the Library in a very wide range of machine configurations should uncover some errors in it (even though it has been tested in three very different environments before it is released for implementation). What is perhaps surprising is that, of all the errors reported in the Library over a two year period, 22 were discovered during the implementation phase (and the corrections immediately communicated to all implementors), compared with only 18 discovered by user sites. (Another 4 were discovered by contributors after their software had been included in the Library.) Therefore thorough testing during implementation significantly improves the reliability of the base version of the Library.

b) Imperfect adaptability of algorithms

Although our parameterization of the machine arithmetic has nearly always provided a satisfactory model for our contributors to work with, occasionally their algorithms have been shown not to adapt as well as possible to some subtler features of machine arithmetic. When one of our routines was tested on the Honeywell Series 60, the results for one problem were only accurate to 2 or 3 significant digits, whereas on other machines they were accurate almost to the precision of the machine. The cause proved to be this: for the Honeywell double precision implementation, the underflow threshold (2^{-129}) is not much smaller than the square of the relative machine precision (2^{-62}). During computation rounding errors introduced a value of about 0.01 x machine precision which in exact arithmetic would have been zero; subsequent computation should have compensated for this perturbation, but the routine tested the square of the offending value: this underflowed, the compensating computation was skipped, and the error festered. Again this was a fault in the base version of the Library which has now been corrected.

However in our experience the unanticipated problem of implementation have more often been caused by deficiencies in the computing configuration (hardware, FORTRAN compiler, link-editor, compiler library) than by deficiencies in the NAG Library - especially as the Library is implemented on an increasing number of mini-computers.

c) Hardware errors

Fortunately, very rare but we did encounter them in the initial stages of implementation on the ICL 2980 and on the Prime 400. They have been corrected.

d) Compiler errors

The most common cause of unanticipated problems; most implementations encounter at least one compiler error. Even with the CDC FTN compiler there has been a dwindling, but still non-empty set of routines which cannot be correctly compiled at the highest level of optimization. There have been similar optimization problems with the DEC 10 compiler. Those problems can be avoided simply by recompiling at a lower level of optimization. Others can only be avoided by modifying the source text, e.g. two types of error on the Burroughs 6700 connected with the use of subroutine-name parameters. On two machines the compiler proved to be so unreliable that the implementation could not be completed until a new version of the compiler became available.

Although our software cannot be guaranteed to exercise all the code in the Library - so that a compiler error might slip through - we do not know of any user site reporting an error in the NAG Library which has been traced to a compiler error. More than one manufacturer has asked to use the Library and test programs to test their compiler.

e) Compiler limitations

Some compilers have failed to compile the Library because of poorly documented restrictions e.g. on the level of nesting of parentheses in expressions, or on the number of subroutine parameters; and in some cases these restrictions have been overcome merely by reconfiguring the compiler.

f) Link-editor errors

(alias loader, binder, consolidator, composer etc.)

Another area of poorly documented restrictions e.g. on the total number of library routines that can be linked in to one main program or on the number of levels of library routines calling other library routines that can be permitted.

g) Errors in the compiler library

The mathematical functions in the compiler library (e.g. SQRT, SIN, EXP) are nearly always taken on trust. Yet if they fail to perform to the required accuracy, they will degrade the performance of the NAG Library routines which call them. On the Prime 400, the test program for calculating eigenvectors by inverse iteration failed to produce an accepted eigenvector: the cause was an insufficiently accurate eigenvalue which in turn was the result of inaccuracies in the DSQRT routine, which Prime have admitted. There is an obvious need for an independent test program to test the compiler library functions.

Thus, in the diagram at the head of the last section, the "feedback loops" from the implementation phase are repeatedly being activated. Implementation problems prompt us sometimes to correct the base version of the Library sometimes to refine the standards that we impose, sometimes to tighten the checks that we make. To cure other problems, modifications to the source text are confined to the implementation concerned, especially when the problems arise from faults in the machine environment - in which case we hope that the modifications will be only temporary until the fault has been corrected.

Finally the fact that this corrective phase in implementation continues to be necessary justifies NAG's policy of always distributing to sites a compiled and tested version of the Library (as well as the source text). Numerical software is not fully developed until it has been shown to perform satisfactorily in each computing configuration in which it is going to be used. Comprehensive testing in each configuration is essential to ensure

a) that the algorithms adapt correctly to the configuration

b) that the configuration is fit to support the software.

9. Conclusions

A methodology for the development of transportable numerical software, has been described. It is evident that the production of high quality numerical software is not an exact science but a branch of software engineering which must take account of the vagaries of a real and changing world. The approach which serves us well now will not necessarily continue to do so.

References

[1] Rice J.R. Software for numerical computation. In Research Directions in Software Technology, P. Wegner (ed), M.I.T. Press (1979).

[2] Rice J.R. Algorithmic progress in solving partial differential equations. SIGNUM Newsletter 11, no. 4, pp 6-10, (1976).

[3] Smith B.T. FORTRAN poisoning and antidotes. In Portability of Numerical Software, Lecture Notes in Computer Science, 57, pp 178-256 (1977).

[4] Ryder B.G. The PFORT Verifier. Software - Practice and Experience, 4, pp 359-377 (1974).

[5] Dorrenbacher J., Paddock D., Wisneski D. and Fosdick L.D. POLISH, a FORTRAN program to edit FORTRAN programs. Dept. of Computer Science, University of Colorado at Boulder, Ref: CU-CS-050-74 (1974).

[6] Du Croz J.J., Hague S.J. and Siemienuich J.L. Aids to portability within the NAG project. In Portability of Numerical Software, Lecture Notes in Computer Science, 57, pp 389-404 (1977).

[7] Cody W.J. The construction of numerical subroutine libraries. SIAM Review, 16, pp 36-46 (1974).

[8] Wisniewski J.A. Some experiments with computing the complex absolute value. SIGNUM Newsletter, 13, no. 1, pp 11-12 (1978).

[9] Reinsch C. Some side effects of striving for portability. In Portability of Numerical Software, Lecture Notes in Computer Science, 57, pp 3-19 (1977).

[10] Ford B. Parameterization of the environment for transportable numerical software. ACM Trans. Math. Software, 4, pp 100-103 (1978).

[11] Ford B., Bentley J., Du Croz J.J. and Hague S.J. The NAG Library 'Machine'. Software - Practice and Experience, 9, pp 56-72 (1979).

[12] Du Croz J.J. and Fosdick L.D. Incorporating a suite of routines into a library: a case history. In preparation (1979).

[13] Hague S.J. and Ford B. Portability - prediction and correction. Software - Practice and Expereince, 6, pp 61-69 (1976).

[14] Wilkinson J.H. Rounding Errors in Algebraic Processes. Notes on Applied Science, No. 32, Her Majesty's Stationery Office, London (1963).

[15] Schonfelder J.L. The Production of special functions for a multi-machine library. Software - Practice and Experience, 6, pp 71-82 (1976).

[16] Boyle J.M., Frantz M.E. and Kerns B. Automated Program Realisations: BLA Replacement and Complex to Real Transformations for LINPACK, Argonne National Laboratory Technical Report (in preparation).

[17] Brown W.S. 'Software Portability', NATO Summerschool (1969).

[18] Cody W.J. Machine Parameters for Numerical Analysis In Portability of Numerical Software, Lecture Notes in Computer Science, 57, pp 49-67 (1977).

[19] Ford B. Preparing Conventions for Parameters for Transportable Numerical Software, In Portability of Numerical Software, Lecture Notes in Computer Science, 57, pp 68-91 (1977).

[20] Waite W.H. Building a mobile programming system. Computer Journal, 13, No.1, pp 28-31 (1970).

Tools for Numerical Software Engineering

S.J. Hague and B. Ford

1. Introduction

Software tools in the context of this paper are programs designed to assist in the development, testing, implementation, maintenance and distribution of computer software. (These tools are sometimes referred to as programming or mechanical aids). This paper deals with the subject of software tools that operate on FORTRAN software, though equivalent tools exist in some cases for other high-level applications languages. Few non-trivial programs, once completed, remain entirely unaltered throughout their computing life. In the next section we consider why it is necessary to analyse and modify numerical software and the benefit of mechanising such processes. Then we summarise software tools in use by numerical software groups; the experiences of the Numerical Algorithms Group (NAG) in using some of them; and in the final section describe the recently formed Toolpack project.

2. Manipulation and Mechanisation

Before discussing software tools in detail, we must provide answers to two basic questions which might be posed by the 'lay user' or perhaps by the numerical analyst who suspects that his more software-orientated colleagues are attempting to create a major new branch of computer science with no real need for it. These basic questions are:

(i) why is it necessary to manipulate numerical software?

(ii) does the mechanisation of that manipulation process bring significant advantages?

2.1 *Why May Changes Be Required?*

Answering this question poses little difficulty particularly if one has witnessed the development of NAG from a single machine range library project to its present state in which there are implementations of the NAG FORTRAN Library on 28 distinct machine ranges. Within each implementation there may be sub-implementations for particular computing systems, e.g. on the CDC 7600, there are source text differences between the Small Core Memory and Large Core Memory implementations. Similar problems are faced by other groups who are also interested in developing and maintaining high-quality software on many machines. Whether for reasons of uniformity, portibility or refinement, it is frequently necessary to alter a body of source text either on a small scale or throughout an entire suite of programs. Below are summarised a number of reasons which might precipitate such changes:

- correcting a coding error either of an algorithmic or linquistic nature.
- altering the structural property of the text e.g. imposing a certain order on non-executable statements in FORTRAN.
- standardising the appearance of the text.
- standardising nomenclature used e.g. giving the same name to variables having the same function in different program units.
- conducting dynamic analysis of the text (e.g. by planting tracing calls).
- ensuring adherence to declared language standards or subsets thereof.
- changing the operational property of the text (e.g. changing the mode of arithmetic precision).
- coping with the arithmetic, dialect and other differences between computing systems.
- altering similar algorithmic processes and similar software constructs in large collections of programs.

2.2 *What are the Benefits of Mechanisation?*

Several published papers have eloquently argued the case for the mechanised approach to program manipulation. Standish et al [2] discuss the merits of improving and refining programs by means of an interactive program manipulation system. Perhaps more immediately relevant to numerical software, a paper by Boyle et al [3] discussed the advantages of automating multiple program realisations; that is, deriving by mechanical means, several members (realization) of a family of related programs from a proto-type or generalised program.

The main arguments presented for mechanisation usually concern two factors; economy and reliability. If numerous changes are to be repeatedly made to a large body of software, then the use of a mechanical aid offers the prospect of considerable savings in time and effort. Presumably some poor programmer is relieved of performing what would be a tedious and slow manual task and can be employed on some activity with a greater intellectual stimulus.

The fact that the changes are made mechanically means that we can at least expect consistency. It may also be that the mechanical nature of the alterations are amenable to at least an informal (if not formal) proof of correctness for the transformed program. The study of correctness-preserving transformations is an active field of research, on the basis of which, some developed form of a TAMPR-like system may eventually lead to software tools whose operations are demonstrably reliable.

In the light of the experience of the NAG Central Office in using automated aids, our overall view would be that the use of such aids can indeed lead to greater economy and enhanced reliability. We would add two notes of caution, however. The first is that programming projects in general are prone to take longer than expected. In an organisation of limited resources, practical aims and subject to the day-to-day pressures of both academic and commercial life, the decision to undertake the design and implementation of a new software tool should be taken with perhaps more caution than in, say, a research establishment. A second point of concern is that the use of a software tool may lead to over-reliance upon its effectiveness and so to a temptation not to check the output software closely. After a tool has been successfully operational for sometime, complacency can arise. If the tool is applied to data (i.e. programs) which contravene some undocumented assumption made by its developers, then what we must hope for is that the output is unmistakeably wrong even at a casual glance. If such a contravention caused a somewhat obscure malfunction to occur, however, incorrect coding may be generated without it being noticed.

3. Software Tools in use

To indicate the kinds of software tools in use, particularly by numerical software groups, we give below a list of categories into which tools might fall:

- dialect verifiers (e.g. PFORT)
- static analysers
- data flow analysers
- control flow analysers
- program verification tools
- test data generators
- formatters (e.g. POLISH)
- variant extraction } portability aids
- precision changes } portability aids
- value substitution } portability aids
- structuring tools
- language translators
- probe insertors
- syntax-driven transformers

plus general text editors, string processors ...

As examples, we summarise the properties of widely-used tools from two of the above categories:

- PFORT ([4]), produced by Bell Laboratories is a language dialect verifier. It checks a program for conformity to the PFORT (Portable FORTRAN) dialect of ANS-66 FORTRAN, and also generates information about program entities in tabular form.

- POLISH ([5]) is a FORTRAN-tidying program developed by the University of Colorado at Boulder. Its actions are to

  - recalculate statement labels into a specified order,
  - recalculate FORMAT labels into a specified order,
  - adjust spacing in lists, comments and epxressions
  - indent the body of DO-loops
  - terminate each DO-loop with its own CONTINUE
  - move FORMAT statements to the end of a program unit.

Though individual tools such as PFORT and POLISH are useful, well-documented and easy to implement, the overall state of software tools for applications programming is not satisfactory. Apart from the problems of learning of and obtaining these independently-developed tools, the following difficulties may also arise:

- the software tool developed on one system might be difficult to mount on another,

- the tool from an outside source may not be properly maintained,

- such a tool is likely to have a different user-interface from that of others from different sources,

- its effect may not be easily adjustable,

- its actions may be incompatable with other tools,

- it may rely on undocumented assumptions about the input program,

- each tool of any sophistication must perform some analysis of the program being processed. Much of that analysis is common to many tools but each performs it separately.

4. The use of Software Tools in NAG

4.1 *The Background to NAG*

The main aim of the Numerical Algorithms Group (NAG) project is the development and distribution of a numerical and statistical algorithms library. There are three language versions of the NAG Library; Algol 60, Algol 68 and FORTRAN. The most widely-implemented and used of these three is the FORTRAN version , Mark 8 of which ([6]) contains over 460 user-callable routines. These routines plus their auxiliaries and associated test software comprise over 200,000 source text records, and new material will be added at later Marks.

Thus, the overall software management task faced by NAG is the provision of a large and still expanding Library which is available in several languages and on many machine ranges; there are over forty, distinct, compiled and tested implementations of the NAG FORTRAN Library, for example. Manpower for this considerable undertaking is provided by a combination of full-time personnel and voluntary specialists at a number of institutions in the United Kingdom and elsewhere. Most of the full-time personnel work in the NAG Central Office which is responsible for the overall coordination of the project's activities. The main technical task of the Central Office is the verification and standardisation of contributions to the NAG Library, culminating in the assembly of a further Mark (edition) of the Library, which is then passed on for formal certification under various computing systems. The increasing use of mechanical aids by the Central Office in its library assembly and processing work is summarised below. For a more extensive description of the technical functioning of the NAG organisation as a whole, see [7].

Interest in mechanisation began to grow within NAG soon after the project started in 1970. The Library was originally intended for a single machine range (the ICL 1906A) but from 1972, other implementations were launched, and the need for a multi-variant source text management scheme was perceived. This led to the development of the NAG Master Library File System ([8]). With the rapid growth of other machine-range versions, it became apparent that, as well as having the means to store variants, it was even more important to anticipate and minimise differences between implementations, i.e. to remove those variations wherever possible. From its beginning, NAG had intended to use a subset of the ANS-66 standard for its FORTRAN Library but it became clear that adoption of a dialect intersection policy in name only was not enough; conformity needed mechanical verification. The importance of relegating and isolating machine dependencies was appreciated and the desirability of mechanising other, unavoidable changes (such as precision transformation) was also recognised, as were the benefits of a uniform structure and layout of coding throughout the Library.

Thus a major standardisation exercise, called Mark 4.5, was carried out in 1974/5 to make the NAG Library software more portable and more uniform in structure and layout. (At the same time, documentation was revised so that it could support an arbitrary number of implementations.) From Mark 4.5 onwards, the Central Office started to use software tools such as POLISH and PFORT mentioned earlier, and testing aids such as BRNANL ([9]). It also developed a precision transformer, APT, and a standardisation tool, DECS, which uses the output from PFORT to introduce explicit declarations for all variables into a program (and can revise the order of non-executable statements at the same time). The following diagram indicates the major processing steps in the Central Office procedure for software processing. A few explanatory comments may be required:

- the verification stage is primarily linguistic; algorithmic validation has taken place at an earlier stage

- several minor standardisation processes to introduce NAG conventions may be applied between DECS and POLISH

- in the testing phase, the purpose of the test runs on machines X, Y, ... is to gauge the extent to which test results may differ. This is in anticipation of the formal Library implementation activity later on.

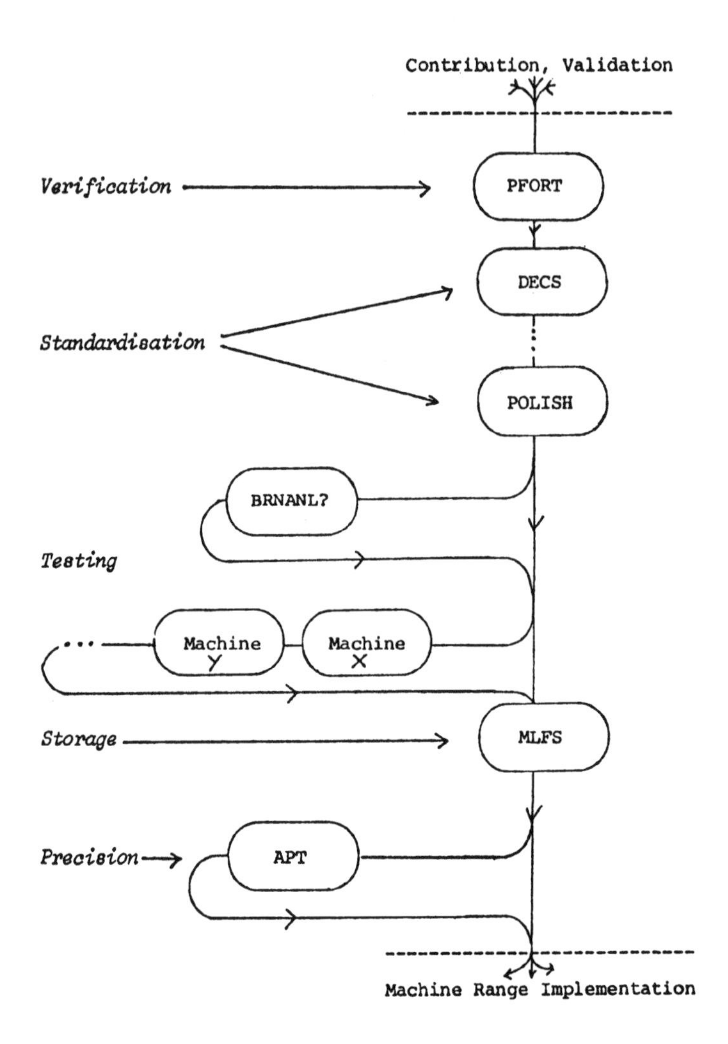

To summarise experience within the NAG Central Office, our view is that without partial mechanisation, the task of Library assembly and processing would be extremely tedious, the code itself would be less consistent, and implementation on different machine ranges much more difficult - a point borne out in practice when pre- and post- Mark 4.5 implementations periods are compared. We are interested in improving our existing tools to make them of use to our own collaborators but also see the need for a more broadly-based and concerted effort to make these and other useful tools more generally available. For this reason, we support the aims of the Toolpack project.

5. The Toolpack Approach (1980)

5.1 *Background to Toolpack*

In the third section of this paper, we drew attention to some of the unsatisfactory aspects of software tools in general. The recently-formed Toolpack project has set out to remedy those defects and may start to have a significant impact in 3 to 4 years time. It is a collaboration effort whose aim, according to the Toolpack prospectus ([10]), is "to produce a systematized collection of software tools to facilitate the development and maintanance of FORTRAN programs." The potential Toolpack user could be any FORTRAN programmer but the proposed facilities are more likely to have immediate appeal to the programming professional or the regular (but non-computer specialist) FORTRAN user engaged in numerical computation.

The reasons for starting a project with the aims and structure of Toolpack can be summarised as follows:

- numerical computation still represents a significant proportion of all computer use, and FORTRAN is by far the most widely used language for numeric applications (and is extensively used in non-numeric areas too). Toolpack is therefore addressing a potentially vast audience,

- software developers such as NAG have come to appreciate the value of software tools both for their own work and for the benefit of computer users generally. There is also some experience of collaborative activities amongst mathematical software groups,

- research into software engineering has matured to the point where an integrated ensemble of tools seems feasible, and the integration attempt would in itself be a worthwhile research effort.

5.2 *Toolpack Development Plan*

The current intention of the Toolpack group is to carry out a three year development plan which, if successful, could lead to the general availability of facilities sometime in 1984. The plan contains three phases, the first of which is already underway (see section 5.3.):

Phase 1: Determination of the functional capabilities to be provided by Toolpack and of the initial packaging and integration environment.

Phase 2: Implementation and documentation of tool components within that environment.

Phase 3: Packaging of the implemented tool components, testing and refinement.

The definition of packaging environment includes consideration of:

- collection of basic tools (such as parsers, lexers, associated table managers).
- input/output primitives,
- specification of the subject language (i.e. the language of the programs upon which Toolpack will operate).
- specification of the implementation language (i.e. the language in which Toolpack tools will be written),
- user/Toolpack interface requirements,
- inter-tool communication conventions.

Investigation of these and other considerations is proceeding. It has already been decided that the principal subject language will be FORTRAN 77 but it may be possible to allow other FORTRAN variants too - on input at least - by concentrating the task of dialect recognition in the front-end lexer/parser rather than in the processing tools themselves. The implementation language has been defined to the extent that whatever language Toolpack participants use initially, it must map into FORTRAN 77. The basis principle of construction is uncontentious, the higher-level tools should be built using lower-level modules where possible. Emphasis on "data-hiding", i.e. not being specific about the exact details of data transfer and structure, is also appropriate, particularly in the initial design stages. A point under active descussion, however, is concerned with the user interface; should an attempt be made to build an integrated information management system which is directive-driven by the user, or should the less ambitious goal of loose confederation of tools be pursued? Under the latter arrangement, the user would drive each tool directly.

5.3 *Candidates for Inclusion in Toolpack*

The following tools are under examination during Phase 1 and may be included in the final set of facilities chosen for release. Other tools not in this list may also be added. In most cases, the tools already exist in a pre-Toolpack form and if chosen for inclusion will be reconstructed according to the construction principle adopted. It should not necessarily be assumed that such existing tools are currently available or suitable for general release. Interested readers are advised to contact the individual developers concerned, via the Toolpack coordinator (see next section).

Tools under review include:

1. DAVE: a static data flow analyser.
   (The University of Colorado)

2. POLISH: a formatter.
   (The University of Colorado)

3. BRNANL: a dynamic test probe inserter.
   (The University of Colorado)

4. BIGMAC: a FORTRAN language macro definer/expander.
   (The University of Colorado)

5. FSCAN/CLEMSW: a tokenizer/parser generator.
   (The University of Colorado)

6. A static semantic analyser/error detector.
   (The University of Colorado)

7. A floating point data generator.
   (The University of California at Santa Barbara)

8. FLOCHK: a structure and programming convention checker.
   (The University of California at Santa Barbara)

9. IPUCHK: an inter-program-unit checker based on the PFORT verifier.
   (The University of California at Santa Barbara)

10. STRUCT: a control flow structor.
    (Bell Laboratories and The University of California at Santa Barbara)

11. The PFORT Verifier: a portability convention checker.
    (Bell Laboratories)

12. The EFL Compiler: a translator for an extended FORTRAN language.
    (Bell Laboratories)

13. FORTLEX: a FORTRAN lexer.
    (Bell Laboratories)

14. APT-X: a precision type converter.
    (The Numerical Algorithms Group)

15. A FORTRAN-Intelligent Editor.
    (The Numerical Algorithms Group)

16. Template Processors.
    (Purdue University)

17. TAMPR: a program transformer.
    (Argonne National Laboratory)

18. A set of basic tools: lexers, parsers, symbol table manipulators etc. (Jet Propulsion Laboratory, Pasadena)

5.4 *General Information about Toolpack*

The participating institutions in the Toolpack project at the present time are:

Argonne National Laboratory
Bell Laboratories
Jet Propulsion Laboratory
International Mathematical and Statistical Libraries Inc.
Numerical Algorithms Group Limited
Purdue University
University of California at Santa Barbara
University of Colorado at Boulder

The project has funding support from the National Science Foundation and the U.S. Department of Energy.

For further information about Toolpack, readers should contact the Toolpack project coordinator:

Dr. Wayne R. Cowell
Applied Mathematics Division (Bldg.21)
Argonne National Laboratory
9700 South Cass Avenue
Argonne
Illinois 60439
U.S.A.
(Telephone: (312) 972 7164).

6. Conclusions

We have described how NAG's own experience in using software tools has led to our involvement in the formation of Toolpack. Our belief is that if Toolpack succeeds in its aims, the programming environment provided for the FORTRAN user will be considerably enhanced. That environment for most users at present consists largely of a FORTRAN compilation system and a general purpose system text editor. Toolpack offers the prospect of a number of additional supporting facilities which will make FORTRAN software easier to develop and maintain.

7. References

[1] HAGUE, S.J.
"Software Tools", In Numerical Software -
Needs and Availability,
Ed. D.A.H. Jacobs, Academic Press, pp 57-79, 1979

[2] STANDISH T.A., KIBLER D.F. and NEIGHBOUR J.M.
"Improving and refining programs by program
manipulation,
Proceedings of ACM annual conference,
Houston, Texas, pp 509-516, 1976

[3] BOYLE J.M. and MATZ M.,
Automating multiple program realisation,
Proceedings of Computer Software Engineering Symposium,
New York., 1976

[4] RYDER, B.G.
The PFORT Verifier, Software Practice and
Experience,
No.4, pp 359-377, 1974

[5] DORRENBACHER J., PADDOCK D., WISNESKI D. and
FOSDICK L.D.
POLISH - A FORTRAN Program to Edit FORTRAN Programs,
Department of Computer Science Report #CU-CS-050-74,
University of Colorado at Boulder, 1974.

[6] The NAG FORTRAN Library Manual, Mark 8
Published (1980) by Numerical Algorithms Group Ltd.,
7 Banbury Road, Oxford, U.K.

[7] FORD B., BENTLEY J., DU CROZ J.J. and HAGUE S.J.,
Preparing the NAG Library,
This Volume.

[8] RICHARDSON M.G. and HAGUE S.J.
The Design and Implementation of the NAG Master
Library File System,
Software - Practice and Experience,
Vol.7 No.1, pp 127-137, 1977.

[9] FOSDICK L.D.
BRNANL - A FORTRAN Program to Identify Basic
Blocks in FORTRAN Programs,
Department of Computer Science Report #CM-CS-040-74,
University of Colorado at Boulder, 1974.

[10] COWELL W.R. and MILLER W.C.
The Toolpack Prospectus
Argonne National Laboratory, Applied Mathematics
Division,
Report TM-341, 1979.

PORTABILITY AND OTHER SOURCE MANAGEMENT PROBLEMS

by

W. Morven Gentleman
University of Waterloo
Waterloo, Ontario, Canada

I. Characterization of the problem

The author of mathematical software rarely has the luxury of composing a program specifically to perform the task immediately before him, i.e. in a specific context, with specific data, on a specific machine. The cost of creating mathematical software is too high, and the number of individuals capable of producing a particular piece of mathematical software is too small, for such directedness to be acceptable. Instead, mathematical software typically has to be portable, so it can readily be made to run in many different machine environments; it has to be modular, so it can be integrated into many other programs; and it must be adaptable, so alternate data structures, algorithms, and other design decisions can be used. An item of mathematical software is thus not a single program, but a family of programs. (Actually in most programming languages the item is often not a program as such, but rather a package of smaller units).

In other words, mathematical software today requires the ability to produce multiple realizations of the same conceptual program. Portability is only one reason, albeit an important one, for doing this. The tools and techniques adequate to handle portability problems are applicable in a broader context, i.e. whenever there is a family of related programs which must be developed and evolved together.

To elaborate these observations, let us begin by looking at the problem of portability of mathematical software. There are several questions we must try to answer.

1. What is mathematical software?

Frequently, in discussions about mathematical software, an item of mathematical software is thought of as the implementation of a single algorithm. The only larger entities considered are libraries, i.e. collections of implementations of often quite unrelated algorithms, presented in a unified and consistent manner.

For our purposes here, we will broaden the definition to include also scientific and engineering codes. These codes are large programs that provide the mechanism to solve any computational problem in some application area. The mathematical theory on which such a code is based is typically quite specific to the application area. Examples of such codes are NASTRAN, a code for solving structural engineering problems; XTAL, a code for crystallographic problems; or GENSTAT, a code for performing statistical analyses.

Including these codes in our definition of mathematical software accentuates the difficulties associated with families of programs because of the sheer complexity of these codes. They often are 20,000 or more lines of source text in a high level language. They often are structured as 250 or more subroutines. They typically have elaborate languages for computational control and data input, combined with flexible options for output and data display - and together these "nonmathematical" parts can make up as much as 80% of the source text for the program. These codes usually interact with stored data structures in the file system of the host environment.

Codes of this kind represent a considerable investment. Often they are built over periods of 10 years or more, so they outlast generations of computers. They contain considerable subject area knowledge beyond the mathematical algorithms. The mathematical algorithms they do contain are often specifically developed for use in these codes. The arguments for the importance of tools and techniques for managing and evolving families of programs are particularly strong with respect to these codes.

2. Whose needs must the tools and techniques address?

In designing tools and techniques for working with families of programs, it is essential to understand how and by whom they will be used, and what aspects will be important. Four situations are immediately apparent.

The first situation is where there are researchers working in the area covered by an item of mathematical software, developing new algorithms. In the initial development stages, they need to perform experiments, modifying the given item of mathematical software to utilize their new methods, or to support detailed instrumentation. The customized versions will probably never be used by anyone except the researcher himself, and typically have a short lifetime. The primary need to be met is that such experimental customization is not just feasible but cheap.

A second situation is that of developers producing or maintaining mathematical software or use by others. Their primary need might appear to be to make the production of new versions cheap, especially versions for other machine environments. Indeed this is important, but more important is the need to make maintenance and simultaneous evolution of all versions possible. Without systematic tools and techniques, the labour of making sure a correction is applied to all versions, of designing an enhancement to be consistent with all versions and ensuring it is installed in all versions, can be prohibitive. Without these sorts of actions, the versions drift apart in quality and maintainability, and soon only that version currently under active development is viable. Audit trails and source code control may be necessary to identify consistent working versions.

A third situation applies to users wishing to tune mathematical software to their own needs. Often the knowledge of preconditions that exist in a particular application can be used to significantly improve the time, space, or accuracy performance of a more general item of mathematical software. The need here is to be able inexpensively to apply the appropriate modifications, or to extract the appropriate subsets of source code, without losing properties of the original. Simple use of a text editor to make the changes, for instance, carries a severe risk of introducing bugs through inadvertently making changes other than those intended.

A fourth situation exists for users wishing to incorporate good mathematical software in their own programs. Here the need is to modify the interface provided by the mathematical software so it meshes with what the program provides. It is not uncommon to find that no single interface fits all needs, and that the designer of the mathematical software cannot anticipate the interface some user of it will require.

Note that in all four situations, the user of the tools and techniques is likely to have considerable sophistication in computing, and to have the motivation to learn to work with tools and techniques even if these are nontrivial, in that the net saving of effort can be substantial.

3. What is portability and why are we interested in it?

A program is defined to be portable if the cost of moving it to a new environment is significantly less than would be the cost of implementing it afresh for that environment.

This definition conceals a number of important issues. First, portability is only meaningful in the context of a set of machines, the "portability set". The cost of porting the program between members of the set will be smaller if the members of the set are more similar, and higher if they are more dissimilar. Hopefully the portability set will be characterized by properties its members must exhibit, rather than by a list of machine models and operating system and language processor release numbers. The universal set, of all software systems on all machine ranges, is rarely useful.

Second, performance of the ported program does matter. That is, the cost assessed in the definition above should reflect the relative difference in time, space, and accuracy performance between a ported program and a specially written program, over the life of the program. One of the encouraging results of research into portability is that, through the use of appropriate techniques, programs can often be built which are portable and which do not suffer performance disadvantages in production vis-a-vis programs specially built for each environment.

Third, portability is a matter of degree. It is sometimes possible to produce software which is machine independent over some portability set, i.e. where no changes at all are required to port the software to any member of the set. Often, however, this is done by methods which incur severe performance penalties, such as by using interpretive execution. The resulting program may thus not be as portable, in our definition, as one for which some changes must be made to facilitate better performance in different environments. Also, allowing some changes to be made can significantly increase the useful portability set over that which would be possible with strict machine independence (for instance when I/O environments are different).

Fourth, the costs and benefits of portability should not be assessed just at the time a program is ported to a new environment. It is widely recognized that maintenance and evolution costs are a large part of the cost of any program - three quarters of the total cost according to many authors. Most bugs found in a portable program will be present in all versions, so the cost of finding and repairing such errors can be amortized across all versions provided the fix is retrofitted to them all. Clearly this economy of scale does not occur when the program is specially written for each environment. Similarly, most evolutionary changes in a portable program can be independent of changes required to particularize the program to each environment, so if machinery exists and is used to retrofit evolutionary advances to all versions, an economy of scale can be achieved over the cost of having to evolve each version separately. (The latter often leads to having to port the program again at a later date).

Fifth, some of the costs which should be included are the costs of training programmers to work on or with the program. A portable program achieves an economy of scale here, because apart from the identified machine specific aspects of the different versions, all versions appear and behave the same. By contrast if programs were specially implemented for each environment, their internal structure and their external interfaces would almost certainly differ in ways immaterial to meeting their functional requirements, but which would constitute a significant learning cost to people working on or with more than one of them.

Finally, the portable program may become available in the new environment significantly sooner than one specially written for that environment, and this can affect the cost in more than just the implementation labour saved. In both the commercial and scientific world timeliness can be of great intrinsic value, and in particular being available first can be more important than being better but too late.

This discussion has centred on cost, in its various aspects, and leaves the impression that reducing cost is the principal motivation for portability in programming. This is true, but not the only reason it is desirable. Another reason is to guarantee comparability. In general it is very difficult to compare machines, or algorithms, or problems, because so many factors unrelated to the comparison of interest are also different, in unknown ways. If a portable program is used to solve the problem, many potentially different factors that are irrelevant will actually be done the same way, and those that are irrelevant to the comparison but done differently are known and identifiable so their effect can be assessed.

Another reason portability is desirable is that it can guarantee preservation of properties of the portable program. This may be useful because proofs of these properties may not need to be separately performed for each version. It may be useful because a user can count on these properties in integrating the program into his application. It may be useful in that performance may be expressible in terms of these invariants. It may be useful just in terms of not forcing the user, given each version, having to establish what properties to expect this program to have.

4. In what ways can mathematical software not be portable?

Since it has been demonstrated for many years that portable software usually need not incur penalties in performance during production execution, there is hardly any explanation for software that is not portable other than that the developers did not consider ramifications for portability of the way the software was implemented. There are several ways in which software can fail to be portable.

a. Language processor dialect differences

Some programming languages are so loosely defined that dialects appear with syntactic and semantic differences. Others are so deficient in functionality that dialects appear with incompatible extensions. These dialect differences are more of a nuisance than a serious problem, in that either source text must be restricted to be in the intersection of the dialects, or a preprocessor must be provided to map the real source for the software to whatever the target dialect requires. The principal difficulty is identifying the common subset and ensuring programs are confined to it. Work on standards [3,8,15,51,67] and portability verifiers [22] in recent years has alleviated this difficulty.

b. Properties of floating point arithmetic

It once was felt that the great difference between floating point arithmetic systems (to say nothing of the glaring deficiencies in some) made mathematical software intrinsically nonportable. Accuracy and range characteristics can affect the domain of problems that can be solved, and even the algorithm of choice to solve them. Axioms satisfied by the floating point arithmetic system can be essential to proving a method works or is efficient. Recently [34,43,65,66] it has been recognized that properties of any floating point arithmetic system can be described by a model with relatively few parameters, parameters that in practice have limited choices of values. Programs expressed in terms of these values are portable.

Surprisingly often these parameters merely determine the values of constants appearing in expressions in the software. Sometimes, however, they must be used to select the number of iterations to be taken, the order of approximation to be used, or even which of several algorithms is appropriate.

c. Range of integers

Fortran programmers, familiar that the Fortran standard requires integers and reals to occupy the same size storage unit, and aware that useful floating point representations require at least 32 bits, are often surprised to find portability problems arising from inadequate integer range. The problem is common with minicomputers, but can happen with large machines too. Many language implementers assume integers are only used as array indices, and so choose representations, registers, and machine instructions that do this efficiently. This can result in overflow for more general computations which were done with integers for exactness, or even just for counters that can take on large values.

d. Pointer range and memory size

An obvious constraint on the portability of programs is that they must have enough memory available to satisfy their requirements. However, there are two additional complications.

The memory requirements of many programs are data dependent, so the effect of smaller available memory is merely that larger problems cannot be solved. Statically demanding the maximum memory such a program could ever use restricts the portability unnecessarily as well as, in many computing centres, unnecessarily increasing the cost of running the program. The difficulty is how to make a data dependent amount of memory available to the program. The problem can exist on two levels: how to change the amount of virtual memory provided to this program by the operating system, and how the program can subdivide and use this memory. In some languages, and some operating systems, dynamic memory allocation is the norm, and there is no problem. Other languages, such as Pascal, allow memory to be allocated dynamically from a stack or heap, but the total amount available for allocation is often fixed by JCL when the program is run. Fortran, of course, fixes both. The typical solution in Fortran subroutine packages is to make the user provide data and work areas. This solution is neither convenient nor possible in all cases, and imposes an unreasonable burden on the user. Consequently many large codes in Fortran implement their own dynamic storage manager. One problem with this is that the size of the storage pool then appears as an array dimension in many subprograms, and according to the Fortran standard the size of the pool thus cannot even be changed by JCL without recompiling, let alone change dynamically. Of course many Fortran systems allow this problem to be solved in a nonstandard way by putting the pool in blank common and allowing references at run time past the declared size. The actual size available for blank common is determined by the JCL to run the job and is made available somehow, and sometimes the size of blank common can be grown at run time by system call when required. The portability problems here are obvious.

A different portability problem which a storage manager within a program can give rise to, but which can also arise in its own right, is the problem of pointer range. Many machines today have short pointers, say 16 bits. The effect is that no array can have more than a fixed number of elements, say 64K, even if the program may have many such arrays. Typically this maximum sized array is too small for the storage pool described earlier. Certain other programs, such as programs for spectral analyses, occasionally also find the maximum sized array too small.

e. Demand paged versus real memory

Although demand paging is intended to give the user the impression of having a vast one level store, the performance of programs can be severely degraded if the pattern of memory references does not exhibit locality. This means programs for demand paged machines may need considerable reorganization from what would be natural for execution on real memory machines. A simple example is that inner loops may need to scan down columns of matrices rather than across rows, and this requires expressing some matrix algorithms in an unfamiliar form. A more striking effect is where extra computation may be worthwhile to avoid the overheads incurred by scattered data references. This occurs, for instance, with the FFT.

f. Vectorization, attached processors, and compiler optimizations

Recognizing that certain types of computation are prevalent and extremely important in particular applications of computation, hardware and software designers have produced systems optimized to these needs. For example, add-on array processors are popular for minicomputers in certain applications. Designing mathematical software to exploit these optimizations if they are present, and yet to have satisfactory performance when they are not, can be challenging.

g. Assumed available library

Although a programming language is the raw material from which programs are made, the tools and standard components that permit the efficient production of software are the elements of the library available to the programmer. The libraries defined as part of the standard languages are very primitive. For building mathematical software, one would like to assume numerical components at the level of the NAG library or the IMSL library. Unfortunately these two do not quite contain functional equivalents, let alone common user interfaces, and many sites have neither. Consequently the designer requiring a service that should be met from the library must provide his own, or accommodate the variations of module with which he may be presented.

If this problem is bad for numerical components, it is far worse for non numerical components of the library. As an example, the language C, which is one of the most successful languages for writing portable software, has two rich but incompatible libraries: the Unix library and Whitesmiths library.

h. The I/O interface

Algol 60 tried to ignore it, and for implementation of small numerical algorithms it is often unimportant, but in larger units of mathematical software the way code interfaces to the user is often crucial. Data must be input for each run, databases of permanently stored items must be referenced, control directives must be accepted from the user. Output must be produced, both for human perusal and subsequent machine processing. Results are often only meaningful to humans when displayed graphically. Unfortunately there is little agreement on what should be available and how access to it should be structured, never mind a standard for the interface. Thus a significant part of the scientific and engineering codes described earlier is often devoted to graphic input and output; menu input, data or control languages and data screening; and file system and data base access. All this typically represents a portability problem.

i. Operating system interface

In our experience, however, the hardest portability problems are associated with requesting services from the operating system. The available services differ on every system, as do the way these services must be requested. Resource acquisition, e.g. increasing the allocated virtual memory, has already been mentioned. How to achieve interaction between a user at a terminal and his running program differs. Exception handling is important in mathematical software, and not is even vaguely standardized. Statistics logging, if available at all, is different on each system. JCL specification, required if the popular preprocessor approach to portability is to be used, is far from standardized despite some noteworthy efforts to compile machine specific JCL from a machine independent form [26,27,28]. These problems have lead to research in portable operating systems, which present the application program with the same interface on each machine [55,57,58].

6. Representations

In trying to build mathematical software in light of the above points, we realize that what we are concerned with is representations of the program. At some abstract level in the programmer's mind, there is only a single conceptual algorithm. All else is implementation detail.

However as soon as the programmer gives this conceptual algorithm a concrete representation as source text in some programming language, the possibility of alternative versions appears. Trivial alternatives are associated with differences in notation as to how the algorithm is expressed. Deeper alternatives are associated with the way certain conceptual computations are achieved, the order in which subcomputations are performed, or the data structures used to represent conceptual entities.

These alternatives typically affect a program not in one place but many, even though the differences are small and systematic.

7. Single source

We stress the importance of the fact that where differences do not exist, there should be only one master copy of the source. This means that fixes and improvements to the version independent code apply immediately to all versions. It also helps the maintenance programmer, charged with changing one version, to locate and consider what other versions may be affected by his change. Conversely, the reader learning the program has a big advantage in that he can identify these parts which do not change between versions.

8. Other reasons for alternate versions

Because of the factors discussed above, portability over a broad set of environments implies multiple versions of a program. But alternative versions do not just arise from portability exercises. There are other problems giving rise to alternate versions, or equivalently, to families of related problems that share much common code,

- instrumented versions of codes facilitate the study of what, without the instrumentation, would be a production code,

-the assumption of different mathematical properties for the data can lead to many different versions of the same program,

-sometimes even if the mathematical properties of the data are fixed, different space/time tradeoffs are possible in the algorithm, leading to different versions of the program.

An interesting example here is Sande's FFT package. Although at some conceptual level there is only one algorithm being implemented, the package is roughly 5000 lines of Fortran, in 75 subroutines, and yet covers only about 1/4 of the variations that would be needed for completeness.

II. Machine Independence and the Choice of Abstract Machine

Program portability has long been recognized as an important objective, and many techniques to attain it were suggested and studied before the unifying idea was recognized. That idea is that there is only one way to achieve program portability: to design the program for an abstract machine, and then to obtain realizations of the program by implementing the abstract machine for each environment in the portability set.

The purpose of the abstract machine is to hide irrelevant detail and to make explicit the detail which matters. It provides an interface which specifies what the portable program can and does assume about the environment in which it runs, and how operations in this environment are described. When programming the portable program, this interface can be used without worrying about how it will be implemented in specific environments. When implementing the interface for some environment in the portability set, the implementor can concentrate on what needs to be supported, without getting into how the portable program will use these things. Different approaches to program portability correspond to differences in abstract machines and the way they are specified or implemented.

An abstract machine can be specified with various different degrees of precision. At the loosest, all that is defined is some of the data structures and some of the operations on them. At the tightest, the syntax and semantics expressing all actions of the abstract machine are rigourously defined. Looser specifications, typified by the MUSS project [58,59] or by Martin Richards' BCPL compilers [13], allow the description of designs which are portable over a wide range of computing environments. However, considerable manual effort is required to port each individual program, as it must be recoded for each environment. Tighter specifications, typified by programming language standards, [3,8,15,51,67] attempt to attain economy of scale so that once the abstract machine is implemented once for each environment, realizations of portable programs can be obtained automatically. The tighter specifications, however, run considerable risk of inadequacy for many programs (as happened with input and output for PASCAL) or of overspecification which unintentionally and unnecessarily restricts the portability set by requiring implementations which are inefficient in certain environments. Note that this can happen despite the designer's expressed intentions, as when the stack organization and direction of stack growth for the language C was determined by the code for the function printf, despite warnings that the code was not portable [54].

An abstract machine can also be specified with various degrees of formality. At the most informal, actions and data structures must be inferred from the way they are used. At the most formal, a notation (the Vienna Definition Language and Bell and Newell's ISP are two very different examples) is used to specify every bit of state the machine must have, and how each bit is affected by each available action. It is often the case that tightness of specification is correlated with formality of specification, although such a relationship is not necessary.

An abstract machine can be defined at any of various different levels of language. There are low level abstract machines, such as IBM 370 machine language or PASCAL P-code. There are high level abstract machines, such as FORTRAN, PASCAL, or ALGOL 68. And there are very high level abstract machines, such as APL or the symbolic algebra languages. The language level characterizes how much the abstract machine is aware of what it is being used for. Higher level abstractions imply both shorter programs and the possibility of efficient implementations over a larger portability set but also imply that the implementation of the abstract machine will be more work than for lower level abstractions. On the other hand, expressing high level program concepts for a low level abstract machine may result in so much overspecification that only the assumption of very sophisticated optimization can assure a meaningful portability set, and even this may not be enough to restore the clear expression of problem which has become cluttered with irrelevant detail. In the author's experience, the higher the level the better.

Once the abstract machine has been defined, various choices are open for how it can be implemented in different environments. At one extreme, hardware may be used that can execute the operations of the abstract machine directly. The microcode or emulation approach is for hardware to interpret the operations of the abstract machine to execute on some simpler abstract machine. Moving across the spectrum, another choice is for software to interpret the operations, at run time, to execute on some simpler abstract machine. At the other extreme, software may translate the operations, prior to run time, to execute on some simpler abstract machine. This extreme of software translation prior to run time itself allows for many variations. The translation may not be fully automatic, requiring instead some manual assistance. The translation may be done in several stages, in which case each intermediate language actually defines an intermediate abstract machine. The translation may assume abstract data types, where issues of representation and algorithmic operation are layered, and textually remote from use. Alternately, it may assume that all possible versions of the code are included inline, and only selection is required. Finally, some software implementations regard the representation of the source text for the program as an object upon which extensive and sophisticated computations are required.

The term "portability set" occurs in the foregoing discussion of abstract machines. The set of hardware and software environments in which it is possible to produce an efficient implementation of an abstract machine is the portability set of that machine. Obviously the portability set is limited by properties specified for the abstract machine. The design of abstract machines is thus closely bound up both with an understanding of the requirements of the program or programs which it will be used to port, and with an understanding of the environments in which implementations may be desired. Often portability failures in the past can be traced to properties unnecessarily or inappropriately specified for the abstract machine due to an inadequate understanding of the potential portability set. Design of a new abstract machine for each portable program, as has been advocated, thus calls for more sophistication and experience than most programmers have.

Thinking about programs for abstract machines and implementations of abstract machines provides a methodology for producing portable programs. This methodology is more useful if properties of the program, such as correctness and performance, can be derived in terms of the abstract machine. This is facilitated if properties of the abstract machine are described parametrically, e.g. base, range, and representational error for floating point numbers. In using these properties, sometimes the parameters can be left open, but sometimes implementation or analysis is simplified if it is known that the parameter lies in a certain range, or has one of a few specific values. Another way in which the methodology can be more useful is if the same, or similar, abstract machines can be used for many programs so that the cost of implementing the abstract machine can be amortized over many programs. There is a tradeoff here, for different programs would allow different aspects of the abstract machine to be unspecified, and so using a common abstract machine can restrict the portability set.

Reference was made earlier to the idea that translation in several stages effectively defines a hierarchy of abstract machines. This idea has actually been proposed on its own - that portability should be obtained by having a hierarchy of abstract machines, each implemented in terms of simpler ones [17,21,24]. No convincing examples have been demonstrated, however, with more than two or at most three levels in the hierarchy. A different wasy to think of a hierarchy of abstract machines is to have portable programs generally rely on a very high level abstract machine, but have the translator that expresses the operations of this very high level abstraction in terms of any of several low level abstract machines be itself a portable program, written in terms of an abstract machine especially designed for writing such translators. In this way the cost of producing specialized and efficient translators could be kept down.

At this point it is appropriate to introduce some definitions with regard to the idea of machine independence, as opposed to the weaker idea of portability. A program is said to be machine independent for a certain abstract machine if there is a single machine readable representation of the program for the abstract machine. A program is machine independent over some portability set if it is machine independent for the corresponding abstract machine, and if there is a mechanized scheme with no manual intervention for realizing the abstract machine on each of the members of the portability set. This pair of definitions embodies the essence of the loosely stated idea that "a machine independent program is one that runs on other machines without change" while avoiding the internal inconsistencies of taking the loose statement literally. Few programs of any size or complexity can be machine independent over portability sets that are at all large, and as noted before, a machine independent program may not be as portable as one that is not machine independent, because of implicit run time inefficiencies.

III. Classical Approaches to the Problem

In this section we will examine the classical approaches to programs that must exist in several versions, such as is implied by portability. Actually our survey will be more restricted than that. In many cases where software can be configured in several different versions, there is a question of when this configuration is done. Is it done prior to run time, so there is a "generation step" which produces the version desired, which is then run? Is it done during an initialization phase at run time, in that the code for all versions of the program is loaded at run time, then interactive or direct hardware enquiry determines the particular version desired, and the unnecessary code is released before computation begins? Or is it done as required, in that the code for all versions is loaded at run time, and flags are interrogated at each decision point to see what version to use next? There are attractions to the dynamic configuration, but often it is not feasible at all, and even when it is, it may not be practical for reasons such as core requirements and computer charging formulae. Our survey will be restricted to the techniques for particularization prior to run time. Note that this coincides with software translation prior to run time which was discussed in the last section as the most popular way to implement abstract machines. Whatever way a program is designed for porting, diagnostics should be designed with it to confirm that the port was done correctly.

1. Numeric Manifests

The simplest case of producing multiple versions of programs is where the differences between versions can be avoided by naming literal constants, using the names wherever the constants are required in the code, and providing the appropriate list of constant values separately. The constant values can be supplied manually, or can sometimes be obtained by simple computational experiments in the target environment. Many necessary constant values can be derived from others once they are known. It is very desirable that test diagnostics are provided to confirm by consistency and plausibility checks that the constant values are specified correctly. The surprising thing about this approach is that, when used in conjunction with a standard-conforming language processor, it handles a wide range of portability problems or other situations requiring multiple versions of programs. IMSL, the Bell Labs PORT library, and IFIP Working Group 2.4, among others, have published lists of constant values that handle the great majority of portability problems in numerical computation [4,12,31,43,52]. (Some of these "environmental inquiry" parameters can equally well be provided by function call at run time.)

Manifest processing is part of the language definition for many programming languages, but can readily be added by a simple preprocessor when it is not. There have been suggestions that the "environmental inquiry" manifests should also be incorporated into language standards so users would not be responsible for having to supply values.

We note in passing that there is some difficulty in grouping value definitions together in order to reflect the structure of the source code, while still grouping together values which are machine dependent and apply to a specific machine. Such problems in organizing manifest declarations might be ignorable when there are only a few dozen manifests, but become a substantial practical problem when, as in one system we are aware of, a complete listing of manifest definitions takes 130 pages.

2. String Substitution Manifests

The manifest technique can be made more powerful (and is, for example, in the programming language C [54]) by what is apparently a small change in semantics. Instead of the manifest being a name for a literal constant, it is the name for a textual string which is, in effect, substituted into the source code before the source is pursed by the compiler. Wherever the simpler numeric manifests can be used, this is identical in appearance and effect, although the implementation is necessarily less efficient. The advantage comes, however, in the observation that other changes can now be made in the source code, beyond the values of numeric constants.

One place where it can be used to good effect is to change partial syntactic structures in the program. For example, a manifest called FLOAT could be defined for use in declarations, so the value REAL or DOUBLE PRECISION could be supplied as appropriate for that machine.

Operating system dependencies, such as how to create and open a file, or language extensions, such as how to reread a record with a different format, are other examples where a string substitution manifest can remove machine dependencies from the general source into the manifest definitions.

An important use can be to conditionally delete code. For example, if a certain function call might not be needed on some realizations, the function call might be replaced by a manifest that would either have as value the function call, or a null string, as appropriate. This same technique is often useful for deleting local variables which are not needed in some realizations, or both actual and formal arguments to procedures when some realizations do not require them.

3. Conditional Compilation

The idea of conditionally deleting code forms the next major technique used for portability and other situations where programs must exist in several versions. With this technique, the setting of switches dictates whether a block of source code will be deleted before the source is passed to a compiler. Sometimes the syntax used is extended to specify which of two, or which of several, blocks of code is not to be deleted: sort of a precompilation if-then-else or case statement. Note that the syntax must be distinct from the run time if-then-else or case statement, and cannot just be the run time statements optimized at compile time to remove dead code. We will want to use conditional compilation in places where the run time conditionals cannot be used, such as to delete declarations or arguments as discussed above under string manifests. The various alternatives of a conditional compilation case statement may well be incompatible and duplicated declarations or statement ports, when considered altogether as part of the language.

The way this technique is used to implement multiple realizations of source code is evident: all possible versions are written together in one master version of the source, and by appropriate switch settings all but the appropriate code will be deleted for each version desired. If the syntax for conditional compilation is defined appropriately, it may be that one of the versions will be directly compilable even if the conditional compilation pre-processing is not done, the other versions being treated as comments. In the case where conditional compilation is implemented by a separate preprocessor and is not part of the standard language, this has been claimed to be a major advantage [31].

Conditional compilation is clearly a complete solution to the multiple version problem, but it is not a satisfactory one. There are three interrelated reasons for this. Code full of conditional directives is clumsy and difficult to read; locating all places affected by a switch is difficult because they are not localized; and there is no assistance in deriving the additional code required to port the program to a totally new environment. Because conditional compilation is often advocated as the total solution to all portability and version problems, especially by novices at porting or those familiar only with small problems [31], we will elaborate these points.

There are two intrinsic reasons why code that has been made portable this way is difficult to read. One is that as described above, adequate conditional compilation must be done with a syntax that is recognizably different from that of the run time language. Usually this is done by a separate lexical structure, or perhaps by some syntax related to the physical organization of the source into lines, columns, pages, files, etc. This structure clashes with the syntactic structure of the program: the same source must simultaneously be understood at two different levels. For most programming languages the run time language has a smooth and constant flavour, but the conditional compilation language is cruder and jars with the constructs of the run time language.

The other reason is less aesthetic and more a simple statistical observation. Suppose a portable program requires only 5% of its source statements to be changed for every different environment, but the portability set over which the program is to be portable involves 40 different environments. The single master copy then contains twice as much machine specific code as machine independent code! Finding, amongst all the lines of source, those which are machine independent or those which apply to a particular machine environment (machine specific and machine independent) is hard, and trying to follow the control flow, for instance, is exceedingly awkward.

There are three intrinsic reasons why it is difficult to locate all places affected by a switch defining a specific version. The first is simply that version specific code is only identified by the appearance of the conditional compilation syntax in the source code. The more successful we are at avoiding unnecessary version differences, the higher the proportion of code which must be searched to find the version specific parts. If the task is to understand the differences between versions, as it is when a new version is to be developed or a major design change introduced, wading through tens of thousands of lines of irrelevant code is inconvenient, to say the least.

To reduce the sheer mass of machine specific code, it is desirable that the switches that define which code is to be selected not be directly the switches that define the choices of realization that are meaningful, but be computed from them (or vice versa). In this way, whenever some of the versions can share version specific code, this code need not be replicated. However now the task of locating, for the reader, all code applying to a particular version is complicated by having to evaluate these functional dependencies amongst the switches, and there may be little indication of why one sharing occurs and not another.

The third reason why it may be difficult to locate all places affected by a switch is that the source code is organized by program structure, and not by the data abstraction implemented, or by the environment realized. This means that the attribute defining a specific version affects not just one localized place in the code, but many scattered places, even if the dependencies at each place are conceptually similar.

The difficulties discussed above are particularly evident when trying to augment the master source to include code for an environment not previously considered. However, there is another aspect of the conditional compilation approach which makes this difficulty worse. Typically one finds that, viewing all the alternative code sequences conditional compilation could select, that one is only given specific instances of how something could be done, but no clear indication of the underlying abstraction, much less an automated route to produce the new code sequence for a new environment. This is particularly a problem if the original author is not doing this port. In general in computer science, our tools for going from the abstract to the specific are much better than our tools for abstracting or for identifying isomorphisms. It is much better to have the original programmer describe what is happening at a more abstract level.

4. Macros

Enhancing string substitution manifests to include parameters which are evaluated and passed at substitution time changes these into what is known as macros [30]. By itself, this is a relatively small advantage in a high level language. Macros are heavily used in large assembly language programs, but close inspection shows that this is typically to achieve coding idioms that are directly part of a high level language: generating calling sequences, data structure declarations, control structure implementation, space allocation from stack or freelist, system service requests, etc. In high level languages the enhancement of parameters, by itself, primarily allows the implementation of abstract data types where these are not directly supported in the language, or provides in-line procedure expansion where the compiler does not provide this and the calling sequence is expensive enough to require it.

The real power of macros comes from the fact that most macro processors provide other programming constructs which are executed at substitution time. Substitution time, therefore, becomes a whole new step in the computation, where the code of the program is the object upon which computation is performed.

For example, substitution time expression evaluation gives constant folding, allowing a constant expression anywhere the underlying language requires a constant. (Fortran and Pascal, for example, do not define constant folding as part of the standard language). Substitution time *if* statements and *case* statements give conditional compilation. Substitution time iteration statements facilitate things like unrolling loops (a technique demonstrated to be significant in improving the efficiency of, for instance, the Basic Linear Algebra subroutines [53]). Substitution time function calls, including recursive function calls, allow local representation of global changes (e.g., abstract data types). Macro expansions can sometimes be used to produce additional macro definitions which themselves are used at substitution time.

A basic issue in macro processor design is whether the macros are lexical or syntactic. In principal this is the distinction of whether the macro is recognized by some simple lexical device, or whether the macro is recognized by a template match or more sophisticated syntactic analysis. In practice, the issue is how well the macro (substitution time) language integrates with the underlying run time language. Lexical macro processors are notoriously hard to program and consequently are error prone, but require no integration with the host language. Syntactic macro processors often give up programming functionality in avoiding conflicts with host language syntax, nevertheless the problem of working with programs defined at two levels is at least as bad as alluded to under conditional compilation.

An interesting consequence of the use of macros is that it can influence the specifications of the host language. For example, it has been recognized in recent years that the block structured scope rules, with inherited scope, which were introduced by Algol 60 are not desirable, and many new languages have abandoned them. However, in order that a macro expansion can use a local identifier for a temporary within the macro, and not force the temporary to be passed as a macro parameter nor force the expansion to maintain a symbol table of the entire program and modify the appropriate declaration block, Algol 60 block structure has had to be reintroduced.

A fundamental problem with macros is that the interleaving of text to be modified and directives to modify it makes reading the combination hard even if the use of macros is well-disciplined, and hopeless if it is not. It also makes processing the source with various source code analyzers hard.

5. Editor Scripts

The advantages of macros, without the last-mentioned disadvantage, can be accomplished by removing the editing commands from the source file. Of course markers must be left in the source text, indicating where editing actions should occur, but the specification of what editing action to take can be remote. Indeed this opens the possibility that different editing actions could be applied for different purposes.

Of course macro processors can often be used this way, with discipline, but so can stored scripts for programmable text editors or programs for string manipulation languages like SNOBOL.

A powerful auxiliary feature possible with editor scripts or string manipulation languages is to allow manual intervention. The transformation program can often be simplified enormously if resort can be made to a human to choose which of several transformations applies, or to do a particular transformation manually. Of course this changes the mode of use somewhat. Whereas a macro processor might be applied automatically whenever compilation is to be done, and the substitution time output thrown away, we would prefer to keep manually assisted transformations to keep and only redo them when something changes.

6. Special Purpose Compilers

The macro definitions, together with the macro processor, can be considered as implementing a new language, as obviously can editor scripts together with the editor. Such new languages can also be implemented by "preprocessors". However compiler technology has advanced to where, given access to tools like parser generators, it is quite feasible to write special compilers to perform particular program transformations. The output of such a compiler would normally be the same base language again (possibly augmented), although an equally valid approach is to generate a common intermediate language. In the former case, the special purpose compiler can be considered as just an elaborate editor.

There are several examples of this approach. S. Feldman built a Fortran 77 to C compiler like this [63]. The MCA Vectorizer is a program which has been used at several scientific laboratories to restructure Fortran programs so that inner loops can be effectively performed by vector operations on machines like the Cray. Many of the common compiler optimizations can be performed entirely in terms of source code, and a program to do this has been described in the literature [19]. Any preprocessor defined to extend a standard language, or to instrument programs for performance measurement, is of this form. There are also preprocessors designed to expand small subroutines inline for machines with expensive procedure calls.

7. Functional Decomposition

The foregoing techniques have made it possible to have differences between versions of a program (for portability or other reasons) that occur anywhere. They also allow the expression of those differences to be minimal. However, they do require processing of every line of the source representation of any program with variants. If we use a language which supports separate compilation, sufficiently cheap procedure calls, and adequate flexibility on conformance of formal and actual parameters, most of the advantages can be obtained while restricting variant versions to differing only in code for selected procedures. By providing a mechanism for selecting which versions of various procedures to use, for example by loading from parallel alternate libraries, it may not be necessary to do any processing on that part of the program which does not change between versions, and the cost of the parallel libraries may be amortized over many programs. There are several other advantages. The reader need only learn the interfaces, rather than studying the whole of the code. The interfaces can be thought of as abstract data types. Or we can think of the base language and library interfaces together as defining a higher level abstract machine.

IV. A New Approach

The traditional approaches can all be viewed as attempts to produce a single source representation of the program, written in some superset language which is implemented by a combination of preprocessors, compilers, and parallel libraries, such that automatic processing yields representations for each of the target environments. Recently we and others have been experimenting with a very different point of view. This new point of view is driven by issues of very large programs, called programming "in the large", and starts by questioning the way programs are represented today [18,29,40,61,62,64].

Representation of programs has traditionally been dictated by what was required as input to compilers, especially lexical and syntactic aspects of particular languages and similar issues of programming "in the small". Direct input to a compiler, however, is only one thing which is done with source code, especially for large programs. For example, Fortran is defined [3,51] as a sequence of cards, with text in columns 7 through 72, and Pascal is defined [67] as a sequence of single case lines of arbitrary length. "In the small" this is what must be passed to a compiler, and hence program source code is usually represented in a file system explicitly as this sequence. However, large programs in either language are usually made up of many modules, which often are separately compiled (at least with some of the standard extended Pascal implementations). In this case, both languages require identical declarations to be repeated in different modules, and for consistency and correctness such identical declarations are best achieved by the mechanism of file inclusion, whereby source code represented separately (in another file) is to be inserted at the specified point. Neither Fortran nor Pascal provide for file inclusion. Consequently, the source representation must first be expanded by a preprocessor before being input to the compiler.

As another example, many languages specify the order of declarative statements with respect to other statements, even other declarative statements. Most languages require declaration before use. The Fortran standard does not allow variable declarations, common statements, and equivalence statements to be freely intermixed. These rules simplify the compilation process, and even aid human reading of small programs. However, suppose we wish to treat certain data structures and the procedures that manipulate them as an abstract date type, and represent them together in a file remote from the main source for the program. We might want to do this so that we could readily change the implementation of the abstract data type merely by changing which file was referenced, or we might want to do it so that the same implementation of the abstract date type could be shared amongst several programs, or we might just want to do it so that all aspects of the abstract data type were textually isolated and identifiable. The file inclusion mechanism referred to in the last example gives a way to separate the source code as required. Unfortunately, if we try to include at one point all the source relevant to the abstract data type, and especially if there are several such abstract data types, it may not be possible to satisfy all the ordering rules for declarative statements.

As a final example, there are directives which are needed by processors other than compilers which are used on program source. Formatting processors require directives to control page layout when generating a listing. Instrumentation preprocessors require triggers to identify code segments to measure. Debugging preprocessors may accept auxiliary statements enabled for debugging but disabled for production. An automatic program prover requires assertions to indicate loop invariants, etc. Because of the perceived requirement that program source code be immediately acceptable by a compiler, directives such as these must be hidden where the compiler will ignore them, that is, in comments. This distorts the comment mechanism which was actually intended to guide human reading of the program, and makes human reading more difficult.

1. Structure

The new approach regards this bias towards compilation as a mistake, and instead regards it as important that the representation of a program be dictated by the way a programmer thinks about the program. If this implies some preliminary processing is required on the source representation each time it is to be given to a compiler, so be it. One immediate consequence of the new approach is thus that structural information about a program should be extracted from the main body of source, and instead be intrinsic in the way the representation is stored. A programmer should not need to search hundreds of thousands of lines of source code, for instance, to find those declarative statements that define an abstraction. The representation should be stored such that first access yields what abstractions are implemented and where, with the clutter of the details of declarations and use concealed to some deeper level of access.

Traditional discussions of programming have been based on how to write a program. For example, the concept of stepwise refinement is based on expressing the given task in terms of simpler tasks. But in the situations that this paper has been addressing, there may be no single program, not at any level of abstraction. Instead, there are a family of related programs, which share common parts. These programs must co-exist, and be maintained and evolved together. Consequently it is vital that a programmer working on one version not lose sight of the others, and how his work affects them. The solution adopted by the new approach is to abandon the idea of writing a program, or of keeping the source representation of any single program. Instead, we think in terms of a database of source code, from which any version of interest can be extracted. The primary programming activity becomes one of maintaining and enhancing the database, and defining the extraction rules that specify particular versions is relegated to secondary importance.

In most programming languages, the appropriate unit of granularity in the database, the unit which is stored indivisibly, is the procedure. Sets of manifests and sets of global data declarations are also stored indivisibly. This choice of granularity imposes no serious limitations, because differences between versions at less than the procedure level can still be represented by different procedures. An important aspect of this choice of granularity is that in extracting a program from the database, order does not matter, only what is taken.

We have chosen to work with hierarchical databases, that is, where the organizational structure is a tree. We have found that this is occasionally inadequate. Nevertheless, there are advantages to limiting ourselves to this, with patch-ups only available to handle exceptional circumstances. One advantage is that it imposes a discipline of trying to organize hierarchically wherever possible. Hierarchies are the easiest logical organizations to understand. They also are easily traversed to exhibit the alternatives that exist at any level. Another advantage of restricting ourselves to hierarchical databases is that it permits us to use a standard tree structured file system as the database system.

This means that familiar tools, such as editors and file comparitors, can be used directly on items in the database. This is important in persuading programmers to use the database, as well as reducing the effort to introduce the system.

As an example of such a hierarchical structure, consider a database of source code for statistical analysis of data. At the coarsest level of subdivision, we might wish to group together all code having to do with analyses of time series. The first pathname component for such files would thus be

/time_series/...

The topic of time series analysis can be divided into many subtopics, from spectral estimation to signal prediction to digital filtering. Following the last of these, source code related to it might be grouped under single directory so that pathnames would start

/time_series/filters/...

Filters can be applied in the time domain or the spectral domain, and to raw signal or various data aggregates. One such aggregate is the periodogram, so we might group all filters that apply to it in a directory so they had pathnames starting with

/time_series/filters/periodogram/...

Filters can be designed with various properties, such as low sidelobes, steep shoulders, flat band pass or sharp selectivity, so a wide selection might be available. One of the common filters is the Parzan window. This might be in a file by itself

/time_series/filters/periodogram/parzan_window

On the other hand, since there are several ways this filter can be implemented, this itself might be a directory also, containing code for each possible realization. One class of realizations, efficient and acceptable when low accuracy is required, is based on running sums, each successive filtered estimate being obtained by adding in adding in later periodogram estimates and subtracting off earlier ones. Code for this, as opposed to alternate versions, might be stored under

/time_series/filters/periodogram/parzan_window/low_accuracy

The tree structure of this source is shown in Figure 1.

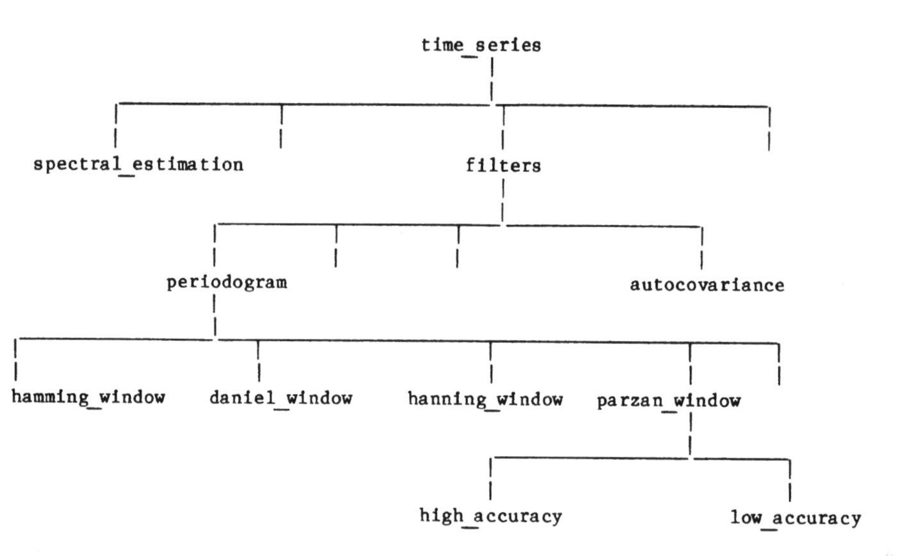

Figure 1. Sample Source Structure

As illustrated by this example, the initial keys in the database structure (components in the file pathname) define the abstraction refinements, and identify the conceptual procedure. Clearly the tree is not of a uniform depth, as various abstractions require different levels of refinement from others.

Once the conceptual procedure is identified, further keys (pathname components) define the specific version of that conceptual procedure. These are used in a specified order. For instance, the foregoing example might need to be refined further by specifying the programming language used. It might need to be refined yet again by specifying the target machine, if the source contained machine specific constructs. The same attribute identifying a version may appear in many subtrees.

Manifest sets and global data declarations are placed in the tree at the lowest level above all procedures where they are used. These too may be subdivided into several versions.

Nodes of special form reference shared subtrees. Other nodes of special form identify test versions, so that development and maintenance can be done within the same database. Backup copies and earlier releases of code are also marked by pathname components of special form.

We have found it useful to store objects other than source code in the same hierarchical database as well. These include command language scripts for building things from source, cross references, listings, object code libraries, documentation, and other things that follow the same hierarchical structure of the source. They can be distinguished from source code by conventions about pathname components.

An interesting observation about programming-in-the-large is that while one might think that top-down design would produce a good hierarchical structure, occasionally it turns out that experience with a body of source shows that a different way of looking at it, a different hierarchical structure, would be better. Fairly massive reorganizations can be done by changing where files are in the file structure, significantly changing how easy the database of source is to work with, and yet not a single line of code gets changed.

2. Tools

Structuring source code in a database is a useful way of organizing it, but in itself would be insufficient if there were not tools provided to work with this database. We have already noted that an advantage of implementing the database with a tree-structured file system is that all standard text oriented tools, from editors to file comparitors, can immediately be applied. Moreover, all standard file system tools can immediately be applied. Listing the contents of directories can show how an abstraction is decomposed or what versions of a procedure exist. Listing system maintained information about a file may show when it was last changed and by whom. Commands to move subtrees from one place in the structure to another can help keep the database organized like the abstraction when the latter changes. Facilities to apply a command to every file in a subtree, or more generally to every file on some list of files ("execute on"), are invaluable for making global updates in the database. Many commands can produce a list of files exhibiting some specified property.

Tools are required, however, that are specific to this file structure being a database of source. The most important is the tool which extracts from the database a collection of source objects to be treated together, especially to be compiled together. Some workers in this area have defined complex configuration languages to specify which objects are to be connected when and how. Instead the tool we use, a program called the inclusion builder, is driven by the structure of the database, which is discovered by traversing it. The inclusion builder is given a list of abstractions and versions to be included or not to be included, and any unresolved questions it interrogates the user to decide. The output of the inclusion builder is a list of files that satisfied the steering abstractions and versions. This list can be input to such programs as compilers or listing generators. The inclusion builder can also update the directive file it is given so as to delete abstractions and versions no longer relevant, and to incorporate decisions made interactively by the user.

Another pair of tools, build_xref and show_xref, provide cross references of symbols. These are more extensive than a compiler would provide, as they encompass all versions, which often could not be compiled together. They also include cross references to symbols appearing in comments, and could include cross references to symbols in documentation. Building such a cross reference is a slow and expensive operation, best performed as a "batch" activity. Fortunately, such a cross reference typically remains valid for long times, so the interactive query program show_xref can provide good response. Show_xref only resolves the cross references to the names of files in which the symbol appears - any finer resolution is done by tools such as editors or "locate in file" commands. The list of files produced by show_xref is often useful input to the "execute on" facility mentioned earlier.

Various listing generator tools can be produced. Although listings are frequently thought of as something compilers produce, they really have quite different requirements that can best be met by separating them from the operation of compilation. For one thing, listings may be wanted of subsets of the source database that are not compilable objects! More importantly, listings are intended for human readers, and should be aware of the possibilities and limitations of printed pages in producing the most readable document form of extracted source. For example, the automatic generation of a table of contents and section headings is invaluable with large programs - say those for which the listing exceeds 50 pages. Cross references by page number also can be useful. With some programming languages, separating the comments out to the facing page or drawing in control structure closures can significantly enhance readability. Given appropriate printing technology, multiple fonts or colour can also be used to enhance readability.

To take advantage of the database structure, it is best if new versions of existing software are developed in place in the database. This is supported by marking test versions by distinguishing pathname components, and providing tools to work with them. The inclusion builder must, of course, be aware of test versions and whether to incorporate them. Other tools required include one to move test versions in as regular versions once development is complete, one to move obsolete versions out to backup status, and finally one to delete versions that are no longer of interest. Since new versions are frequently derived from old, a special editor that predicts the existence of a new version of a file, permits modification (perhaps following a stored editor script), but only saves the file if the prediction is wrong turns out to be quite useful. More generally, the source database can contain master versions of source from which any of the tools in Chapter III of this paper can be used to produce the specific versions desired. This approach facilitates deriving new versions readily, yet allowing for changes in specific versions beyond what the master copy produces.

Finally, we have found that like all large databases, the source structure becomes more complex than what a person can readily keep in mind. Therefore it turns out to be handy to have a tool which, if you do not know quite where to look but can give a partial description of an object, can locate plausible candidates in the database for you. We find it most convenient for this tool to allow you to execute a command in one or more of the suggested candidates, especially to call an editor on them.

3. Final Remarks

The new approach has been found to work very well for Fortran and many other languages. It has, however, intrinsic difficulties with the nested structure and inherited scope of Algol, Pascal, and similar languages. The causes of these difficulties have, however, been criticized for other reasons with respect to "programming in the large" and so perhaps do not represent a shortcoming of the approach.

It might be noted that the new approach is basically a technique for source management. While it is intended to support multiple version software, rather than source code control of evolutionary products, the mechanisms required are quite similar and this machinery is appropriate for that task [20,48].

We have not discussed the distribution and installation of portable programs produced by the new approach. Even with the traditional approaches, these are significant problems [25,39]. The new approach does not lessen the difficulties, especially when commercial realities imply the whole database should not be provided to any single customer.

References

1. J. Strong, J. Wegstein, A. Tritter, J. Olsztyn, O. Mock, T. Steel, "The Problem of Programming Communication with Changing Machines: A Proposed Solution," Comm. ACM, Vol. 1, No. 1 (August 1958), pp. 12-18.

2. T. B. Steel, Jr. "A First Version of UNCOL," Proc. AFIPS WJCC Vol. 19, (1961), p. 371.

3. American National Standard FORTRAN, American National Standards Institute, New York, (1966).

4. P. Naur, "Machine Dependent Programming in Common Languages," BIT Vol. 7, 1967, pp. 123-131.

5. W. M. Waite, "A Language Independent Macro Processor," CACM Vol. 10, (July 1967), pp. 433-440.

6. M. Richards, "BCPL: A Tool For Compiler Writing and System Programming," Proc. AFIPS SJCC, Vol. 34, (1969), pp. 557-566.

7. Orgass, R.J., W.M. Waite, "A Base for a Mobile Programming System," CACM Vol. 12, (September 1969), pp. 507-510.

8. "Clarification of FORTRAN Standards in Initial Progress," CACM Vol. 12, (1969), pp. 289-294.

9. W. M. Waite, "The Mobile Programming System: STAGE2," CACM Vol. 13, (July 1970), pp. 415-421.

10. W. M. Waite, "Building a Mobile Programming System," Computer Journal, Vol. 13, (February 1970), pp. 28-31.

11. P. C. Poole, W. M. Waite, "Input/Output for a Mobile Programming System," Software Engineering Vol. 1, Tou, J.T. (Ed.), Academic Press (1970).

12. K. A. Redish, and W. Ward, "Environmental Enquiries for Numerical Analysis," SIGNUM Newsletter Vol. 6, 1971, pp. 10-15.

13. M. Richards, "The Portability of the BCPL Compiler," Software Practice and Experience, Vol. 1, No. 2, (April 1971), pp. 135-146.

14. A. D. Hall, "The Altran System for Rational Function Manipulation-A Survey," CACM, Vol. 14, No. 8, (August 1971), pp. 517-521.

15. "Clarification of FORTRAN Standards-Second Report," CACM, Vol. 14, (1971), pp. 628-642.

16. R. E. Griswold, The Macro Implementation of Snobol4, W. H. Freeman and Company, San Francisco, (1972).

17. M. C. Newey, P. C. Poole and W. M. Waite, "Abstract Machine Modelling to Produce Portable Software-A Review and Evaluation," Software Practice and Experience, Vol. 2, No. 2, (April 1972), pp. 107-136.

18. S. P. DeJong, "The System Building System," Technical Report RC 4486, Thomas J. Watson Research Center, 1973.

19. P. B. Schneck and E. Angel, "A Fortran Optimizing Compiler," Computer Journal, Vol. 16, No. 4, (November 1973), pp. 322-330.

20. M. J. Rochkind, "The Source Code Control System," IEEE Transactions on Software Engineering, Vol. SE-1, No. 4, December 1974, pp. 364-370.

21. S. S. Coleman, P. C. Poole, and W. M. Waite, "The Mobile Programming System, Janus," Software Practice and Experience, Vol. 4, No. 1, (January 1974), pp. 5-23.

22. B. Ryder, "The PFort Verifier," Software Practice and Experience, Vol. 4, No. 4, (October 1974), pp. 359-377.

23. P. J. Brown, Macro Processors and Techniques for Portable Software, Wiley, New York, (1974).

24. W. M. Waite and P. Poole, "Portability and Adaptability," Lecture Notes in Computer Science #30, Springer-Verlag, (1975).

25. W. M. Waite, "Hints on Distributing Portable Software," Software Practice and Experience, Vol. 5, No. 3, (July 1975), pp. 295-308.

26. D. Rayner, "Recent Developments in Machine-Independent Job Control Languages," Software Practice and Experience, Vol. 5, No. 4, (October 1975), pp. 375-393.

27. G. N. Baird, "Fredette's Operating System Interface Language (FOSIL)," Command Languages, ed. by C. Ungar, North Holland-American Elsevier, (1975).

28. H. Krayl, C. Ungar, Th. Weller, "Portability of JCL Programs," Command Languages, ed. by C. Ungar, North Holland-American Elsevier (1975).

29. F. DeRemer and H. Kron, "Programming-in-the-Large versus Programming-in-the-Small," IEEE Transactions on Software Engineering Vol. SE-2, No. 2, June 1976, pp. 80-86.

30. A. J. Cole, Macro Processors, Cambridge University Press, Cambridge, 1976.

31. T. J. Aird, "The IMSL Fortran Converter: An Approach to Solving Portability Problems, Portability of Numerical Software, Lecture Notes in Computer Science 57, Springer-Verlag, (1976).

32. W. S. Brown and A. D. Hall, "Fortran Portability Via Models and Tools," Portability of Numerical Software, Lecture Notes in Computer Science 57, Springer-Verlag, (1976).

33. C. O. Grosse-Lindemann and H. H. Nagll, "Postlude to a PASCAL-Compiler Bootstrap on a DEC System-10," Software Practice and Experience, Vol. 6, No. 1, (January 1976), pp. 29-42.

34. W. J. Cody, "Machine Parameters for Numerical Analysis," Portability of Numerical Software, Lecture Notes in Computer Science #57, Springer-Verlap, (1976).

35. W. M. Waite, "Intermediate Languages: Current Status," Portability of Numerical Software, Lecture Notes in Computer Science 57, Springer-Verlag, (1976).

36. B. Ford and D. K. Sayers, "Developing a Single Numerical Algorithms Library for Different Machines Ranges," Transactions on Mathematical Software, Vol. 2, No. 2, (June 1976), pp. 115-131.

37. B. W. Kernighan and P. J. Plauger, Software Tools, Addison-Wesley Publishing Company, (1976).

38. W. T. Wyatt, Jr., D. W. Lazier, and D. J. Orser, "A Portable Extended Precision Arithmetic Package and Library with Fortran Precompiler," Transactions on Mathematical Software, Vol. 2, No. 3, (September 1976), pp. 209-231.

39. M. A. Sabin, "Portability-Some Experiences with FORTRAN," Software Practice and Experience, Vol. 6, No. 3, (July 1976), pp. 393-396.

40. D. L. Parnas, "On the Design and Development of Program Families," IEEE Transactions on Software Engineering, Vol. SE-2, No. 1, March 1976, pp. 1-8.

41. P. J. Brown (Ed.), Software Portability, Cambridge University Press, (1977).

42. R. B. K. Dewar and A. P. McCann, "MACRO SPITBOL-a SNOBOL4 Compiler," Software Practice and Experience, Vol. 7, No. 1, (January-February 1977), pp. 95-113.

43. B. Ford, "Parameterization for the Environment for Transportable Numerical Software," Transactions on Mathematical Software, Vol. 4, No. 2, (June 1978), pp. 100-103.

44. O. Lecarme and M.-C. Peyrolle-Thomas, "Self-Compiling Compilers: An Appraisal of Their Implementation and Portability," Software Practice and Experience, Vol. 8, No. 2, (March 1978), pp. 149-170.

45. B. K. Haddon and W. M. Waite, "Experience with the Universal Intermediate Language Janus," Software Practice and Experience, Vol. 8, No. 5, (September 1978), pp. 601-616.

46. R. E. Berry, "Experience with the Pascal P-Compiler," Software Practice and Experience, Vol. 8, No. 5, (September 1978), pp. 617-627.

47. S. C. Johnson and D. M. Richie, "Portability of C Programs and the UNIX System," Bell System Technical Journal, Vol. 57, No. 6, Part 2, (July 1978), pp. 2021-2048.

48. A. L. Glasser, "The Evaluation of a Source Code Control System," Proceedings of the Software Quality and Assurance Workshop, Performance Evaluation Review, Vol. 7, No. 3 & 4 and Software Engineering Notes, Vol. 3, No. 5, (November 1978), pp. 122-125.

49. A. S. Tanenbaum, P. Klint, W. Bohm, "Guidelines for Software Portability," Software Practice and Experience, Vol. 8, No. 6, (November 1978), pp. 681-698.

50. W. H. Josephs, "A Mini-Computer Based Library Control System," Performance Evaluation Review, Vol. 7, No. 3 & 4 and Software Engineering Notes, Vol. 3, No. 5, (November 1978), pp. 126-132.

51. American National Standard Programming Language Fortran, American National Standards Institute, X3.9 (1978).

52. P. A. Fox, A. D. Hall, and N. L. Schreyer, "The Port Mathematical Subroutine Library," Transactions on Mathematical Software, Vol. 4, No. 2, (June 1978), pp. 104-126.

53. C. L. Lawson, R. J. Hanson, D. R. Kineaid, and F. T. Krogh, "Basic Linear Algebra Subprograms for Fortran Usage," Transactions on Mathematical Software, Vol. 5, No. 3, (September 1979), pp. 308-323.

54. B. W. Kernighan and D. M. Ritchie, _The C Programming Language_, Prentice-Hall, Englewood Cliffs, 1978.

55. D. R. Cheriton, M. A. Malcolm, L. S. Melen and G. R. Sager, "Thoth, a Portable Real-Time Operating System," CACM, Vol. 22, No. 2 (February 1979), pp. 105-115.

56. D. Comer, "MAP: A Pascal Macro-Preprocessor for Large Program Development," Software Practice and Experience, Vol. 9, No. 3, (March 1979), pp. 203-209.

57. M. Richards, A. R. Aylward, P. Bond, R. D. Evans, and B. J. Knight, "TRIPOS-A Portable Operating System for Mini-Computers," Software Practice and Experience, Vol. 9, No. 7, (July 1979), pp. 513-526.

58. C. J. Theaker and G. R. Frank, "MUSS-A Portable Operating System," Software Practice and Experience, Vol. 9, No. 8, (August 1979), pp. 633-643.

59. H. Barringer, P. C. Capon, and R. Phillips, "The Portable Compiling Systems of MUSS," Software Practice and Experience, Vol. 9, No. 8, (August 1979), pp. 645-655.

60. G. B. Bonkowski, W. M. Gentleman, M. A. Malcolm, "Porting the Zed Compiler," Proceedings of the SIGPLAN Symposium on Compiler Construction, SIGPLAN Notices, Vol. 14, No. 8, (August 1979), pp. 92-97.

61. T. A. Cargill, A View of Source Text for Diversely Configurable Software, Ph.D. Thesis, University of Waterloo, (1979).

62. L. W. Cooprider,"The Representation of Families of Software Systems," Ph.D. Thesis, Carnegie-Mellon University, (1979).

63. S. I. Feldman, "Implementation of a Portable Fortran 77 Compiler Using Modern Tools," Proceedings pf the SIGPLAN Symposium on Compiler Construction, SIGPLAN Notices, Vol. 14, No. 8, (August 1979).

64. R. F. Brender, "Generation of BLISSes," IEEE Transactions on Software Engineering, Vol. SE-6, No. 6, November 1980, pp. 553-563.

65. W. S. Brown and S. I. Feldman, "Environment Parameters and Basic Functions for Floating-Point Computation," Transactions on Mathematical Software, Vol. 6, No. 4, (December 1980), pp. 510-523.

66. W. S. Brown, "A Simple But Realistic Model of Floating-Point Computation," Computer Science Technical Report No. 83, Bell Telephone Laboratories, November 1980.

67. A. M. Addyman, "A Draft Proposal for Pascal," SIGPLAN Notices, Vol. 15, No. 4, (April 1980), pp. 1-66.

REMARKS ABOUT PERFORMANCE PROFILES

by

J. N. Lyness*
Applied Mathematics Division
Argonne National Laboratory
Argonne, IL 60439

1. Introduction

The term "performance profile" was first assigned its present technical meaning in the field of numerical software in 1976. It was introduced and studied in an attempt to understand some of the results obtained (Kahaner 1971) when evaluating numerical quadrature routines. Since then it has been helpful in designing techniques for evaluating optimization routines (Lyness and Greenwell 1977, Lyness 1979) and it appears that analogous techniques may be applied to evaluate other items of numerical software. It also provides insight to the behavior of programs in which results from one routine are used as input to another routine. This has been termed the "interface problem."

In section 2 of this paper, performance profiles are described in general, and in section 3, an example of a jagged performance profile is given. In sections 4 and 5, their relevance to techniques for software evaluation and to an interface problem is briefly described.

This paper is intended to introduce and illustrate some of the basic ideas in this area. For a detailed description of some of the applications of these ideas, the reader should examine some of the references listed at the end of the paper.

2. The Performance Profile

In concept, the term performance profile is extremely simple.

One requires a one parameter (usually denoted by λ) set of numerical problems, called a problem family for which the exact solution of each problem depends continuously on λ. One requires an item of numerical software, allegedly capable of providing a numerical approximation to the solution of each member of this problem

*This work was supported by the Applied Mathematical Sciences Research Program (KC-04-02) of the Office of Energy Research of the U.S. Department of Energy.

family. A performance profile is a plot of some aspect of the performance of this item of numerical software, the abscissa being λ.

These guidelines are very broad. At present only a narrow subclass of performance profiles have been used by the author to help to understand the behavior of numerical software.

To construct a performance profile we need to state unambiguously a problem family (which depends on λ), a software item which can handle these problems, and the aspect of the performance in which we are interested. I shall now describe a particular choice for these three items.

The problem family comprises a set of quadratures

$$I_x f(x,\lambda) = \int_1^2 f(x,\lambda)dx \qquad 1 \le \lambda \le 2 \tag{2.1}$$

where

$$f(x,\lambda) = 0.1/((x-\lambda)^2 + 0.01) \ . \tag{2.2}$$

Note that the true solution

$$I_x f(x,\lambda) = \arctan(10(2-\lambda))-\arctan(10(1-\lambda)) \tag{2.3}$$

is an analytic function of λ, and so depends continuously on λ as required by the guidelines for a performance profile.

The item of numerical software is a simple routine which returns the N point Gauss-Legendre quadrature rule result, used with N=10, i.e.

$$Q_x f(x,\lambda) = G_{10} f = \sum_{i=1}^{10} \omega_i f(x_i) \ . \tag{2.4}$$

The aspect of the performance in which we are interested is the (unsigned) error made by applying this software, i.e.

$$E(\lambda) = Q_x f(x,\lambda) - I_x f(x,\lambda) \ . \tag{2.5}$$

Since the exact integral can be expressed in terms of standard functions, there would be little difficulty involved in writing a short program to evaluate $E(\lambda)$ for values of λ between 1 and 2 and then plotting $E(\lambda)$ as a function of λ. This plot is a performance profile and is illustrated in figure 2. It is clearly a continuous function of λ, resembling a mildly distorted sine function crossing the λ axis about twenty times having a maximum amplitude of about 0.05.

The numerical information obtained by carrying out this numerical experiment is not unexpected and of little interest. Briefly, for this set of problems, using this quadrature rule, one can expect results roughly equally likely to be overestimates as to be underestimates, each having an error of 0.05 or less. (Since for $\lambda \in [1,2]$, the exact result $I_x f(x,\lambda)$ lies between 1.46 and 2.46, this represents a possible 4% relative error using ten function values.) As mentioned above, this is not an interesting result.

However, there is one somewhat obvious fact abut these results which is significant. <u>This particular performance profile is smooth</u>. In fact there is no need to plot it to see this. We have already noted that $I_x f(x,\lambda)$ is an analytic function of λ. It is trivial to establish that $G_{10}f$ is also an analytic function of λ. It follows immediately that the difference $E(\lambda)$ is also analytic in λ and so the profile is smooth.

At this point, we have described only one performance profile. Naturally we could obtain other performance profiles by carrying out analogous numerical experiments. To start with, we could construct performance profiles, based on the same problem family and the same aspect of performance, namely the error, but using different quadrature algorithms. If we replace $Q_x f(x,\lambda) = G_{10}f$ by some other <u>fixed point</u> quadrature rule, we find results of the same qualitative nature. That is the profile is different, but still smooth. However, we find a different qualitative result when we become more adventurous, and replace the numerical quadrature software item by one of the more recent <u>automatic quadrature routines</u>. A Fortran routine for an AQR might have a calling sequence

```
AQUAD(A,B,EPQUAD,FUN)
```

where EPQUAD is the required accuracy. We might set EPQUAD = 0.1 and proceed exactly as before to plot a performance profile. The only difference is that $Q_x(f,\lambda)$ is no longer $G_{10}f$, but is the result returned by the automatic quadrature routine. When we do this we do not obtain a smooth performance profile. The profile we actually plot is of a type illustrated in figure 3. $E(\lambda)$ is a piecewise continuous curve having many discontinuities. For a few isolated short sections, $|E(\lambda)|$ is unduly large. For other sections, it is consistently small. But by no stretch of the imagination could the profile be described as smooth. In general it is more like a one-dimensional pin cushion. We sum up the result of this numerical experiment by the following statement: <u>Some automatic quadrature routines have jagged performance profiles</u>.

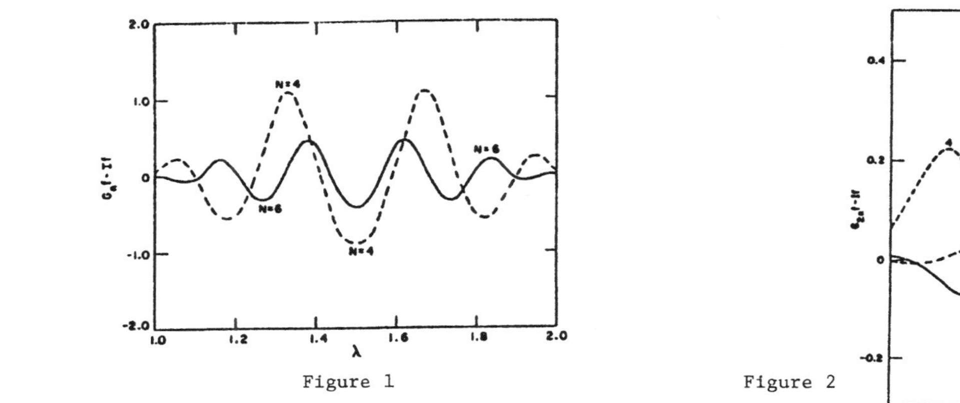

Figure 1

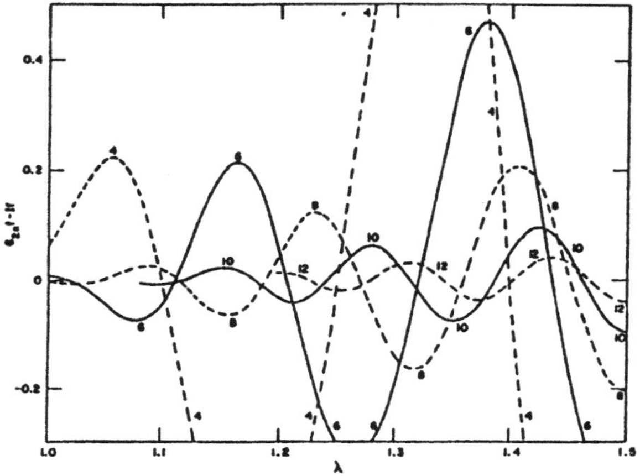

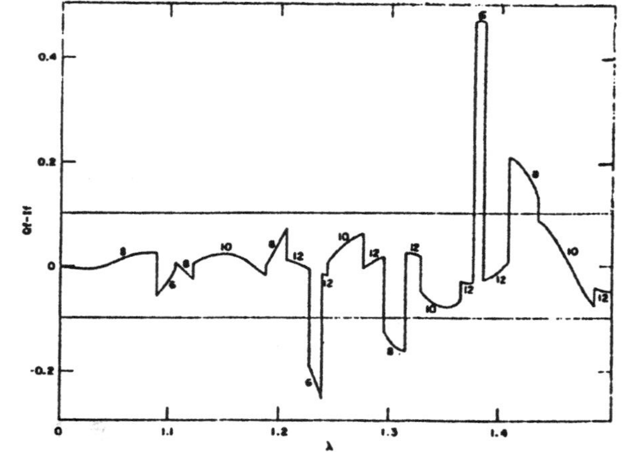

Figure 2

Figure 1, 2, and 3 refer to the problem family

$$If = \int_1^2 f(x;\lambda)dx \quad \text{with} \quad f(x;\lambda) = 0.1/((x-\lambda)^2 + 0.01) \ .$$

In Figure 1, the error curves G_4f-If and G_6f-If are plotted as functions of λ. They intersect six times in the interval $1<\lambda<2$. Thus for six values of λ, $G_4f = G_6f$.

In Figure 2, on an expanded scale, five error curves G_Nf-If, $N=4,6,8,10,12$ are plotted over the range $1<\lambda<1.5$. Each curve is labeled with the corresponding value of N.

In Figure 3, on the same scale as Figure 2, a performance profile is plotted. This corresponds to the above problem family, the naive quadrature routine, and $\varepsilon_{quad} = 0.1$. Each segment of this profile coincides with one of the curves in Figure 2, as indicated.

Figure 3

The reason for this is given in Lyness and Kaganove 1976. However, a brief outline in given in section 3.

We now illustrate another performance profile used in minimization. Descriptions of these may be found in Lyness and Greenwell 1977 and in Lyness 1979.

Minimization routines are complex software items. Their overall purpose is to find an approximation $\tilde{x}_{min}$ to the minimum x_{min} of an objective function $f(x)$. A typical routine requires a starting vector x_0. The routine then evaluates $f(x_0)$ and $g(x_0)$ the gradient of f. It proceeds to make a sequence of function and gradient calls at $x_1, x_2, \ldots$. The location of a particular point x_i is based on the previously calculated function and gradient values $f(x_j)$, $g(x_j)$ $j=1,2,\ldots,i-1$. The strategy is constructed to try to find successively smaller function values and so, in general, with a few exceptions the sequence $f(x_i)$ $i=0,1,2,\ldots$ is a decreasing sequence. The routine terminates under various conditions which depend on tolerances ε, δ and physical limits N provided by the user. One measure of the cost of using a particular minimization routine is the number of function and gradient evaluations it requires to reduce the value of the objective function from h_1 to h_2 where h_1 and h_2 are given values set by the evaluator, satisfying

$$(2.6) \qquad f(x_0) > h_2 > h_1 > f(x_{min}) .$$

Interesting performance profiles may be constructed for testing these routines in the following way. One chooses an objective function $f(x)$ having some interesting topographical feature. One chooses a linear starting region; that is

$$(2.7) \qquad x_0 = a + \lambda(b-a) \qquad 0 \leq \lambda \leq 1$$

where a and b are set so that $f(x_0)$ comfortably satisfies inequality (2.6) for all λ. The resulting trajectory (the set $x_0, x_1, x_2, \ldots$) depends on the value of λ. For each run one records

$$f(x_0), f(x_1), f(x_2) \ldots .$$

One then estimates ν_{12}, essentially the number of function calls required to reduce the function from h_2 to h_1. If $f(x_I)$ is the first function value encountered for which $f(x) < h_2$ and $f(x_{I+h})$ is the first encountered for which $f(x) < h_1$ then

$$n-1 < \nu_{12} < n+1 .$$

Some specified interpolation procedure is required to define an appropriate noninteger measure. Clearly ν_{12} is a function of λ which can be plotted, at least for those values of λ for which x_I and x_{I+h} exist. When an experiment of this type is carried out using any well known minimization routine, it is found that the profile is again highly discontinuous.

These are some special features of the profiles mentioned above which merit special names. The profile in (2.5) may be termed an error-type p.f. since it represents the error made by a numerical process. The one described above may be termed a cost-type p.f. since it measures a component of the cost of a numerical process. Both are termed equal difficulty type profiles as there seems to be no predictable difference in the intrinsic difficulty of the problem when λ is varied. All these descriptors are qualitative in nature. They are not definitions in the scientific sense.

In the software evaluation discussed in my own work, equal difficulty type p.f.'s are used. However, one can visualize perhaps increasing difficulty type p.f.'s in which the difficulty increases with λ. For example, in numerical quadrature a problem family

$$(2.8) \qquad f(x,\lambda) = e^{10\lambda x} \qquad 0 < \lambda < 1$$

would produce error type p.f. in which $E(\lambda)$ takes much larger values near $\lambda=1$ than it does near $\lambda=0$. But the nature of the profiles would remain as before smooth or jagged, respectively.

3. A Particular Jagged Performance Profile

One can envision many types of quadrature routines. Two particular types are:

(1) Rule evaluation quadrature routine

The typical calling sequence is REQUAD(A,B,N,FUN). This approximates $\int_A^B f(x)dx$ by $\sum_{i=1}^{N} w_i f(x_i)$. The user sets A, B and a value of N and the routine uses an N point rule specified in the routine documentation.

(2) Automatic quadrature routine

The typical calling sequence is AQUAD(A,B,EPQUAD,FUN). The user sets A,B and a value of ε_{quad}. The routine uses as many function values as it deems necessary to return a result of accuracy ε_{quad}. It bases its strategy on the numerical values of the function values it encounters. It may use a small number or a large number of function values and it may fail to obtain the desired accuracy giving no warning about this failure.

Many of the properties of automatic quadrature routines are described in detail in Lyness and Kaganove 1976 and in Lyness 1981. In this brief outline, we simply present briefly the reason why the performance profile associated with these routines is usually discontinuous. The reader may find it helpful to refer to figures 1, 2 and 3 when following this description.

A typical automatic routine returns a result of the form

$$\sum_{i=1}^{\nu} w_i f(x_i) \ .$$

However, it may employ different formulas of this type depending on the function values encountered.

A particularly simple automatic quadrature routine is described in Lyness and Kaganove 1976. Faced with members of the problem family described in section 2, when $\varepsilon_{quad} = 0.1$ it returns either $G_6 f$, $G_8 f$, $G_{10} f$ or $G_{12} f$. It calculates first $G_4 f$ and $G_6 f$. If these differ by less than ε_{quad} it returns $G_6 f$ and terminates. When $\varepsilon_{quad} = 0.1$, this happens when λ lies in one of several small ranges, one being $|\lambda-\lambda_0| < \delta$ where $\lambda_0 \simeq 1.380$ and $\delta \simeq 0.05$. Otherwise it calculates $G_8 f$ and if this differs from $G_6 f$ by less than ε_{quad} it returns $G_8 f$ and terminates. Otherwise it continues in this manner. In fact, as mentioned above, in this set of problems for all values of λ it actually returns a result based on one of these four formulas.

The performance profile for this <u>naive quadrature routine</u> then coincides with that of $G_{2r} f$ for those values of λ for which the routine returns $G_{2r} f$. However, as λ is continuously varied, the routine switches abruptly from one Gaussian approximation to another surprisingly often. So the performance profile of this AQR consists of a number of different sections each of which is individually smooth. However, at the values of λ at which a switch from one choice to another is made there is a discontinuity in the performance profile.

Incidentally, it is possible to construct AQRs which have smooth performance profiles. At the present time these are simply theoretical curiosities.

In general it is quite easy to predict whether a performance profile will be smooth or jagged, by examining the algorithm. If this follows different strategies, depending on data received, the profile is likely to be jagged. In cases of doubt, one can always run a short numerical calculation and plot a section of profile.

4. Software Evaluation

The reader will be aware that for many problems in numerical analysis, there exist many different algorithms which carry out the same task but make different use of the available data. For example, as early as 1970 there were dozens of different AQRs available. At that time, considerable effort was expended in carrying out numerical experiments with a view to determining which was "best."

At that time, a testing procedure which came to be known as the battery test was in vogue. A very extensive Battery test of automatic quadrature routines is reported in Kahaner 1971. This investigation involved $N_P(=21)$ different integrand functions and $N_Q(=11)$ different AQRs. Each integral is evaluated by each routine using in turn $N_E(=8)$ different tolerances ε_{quad}. Since the true value of the integral is known in each case, he obtains for each of the $N_P N_Q N_E(=1848)$ runs, three results, namely the accuracy obtained and, the number of function values used and the CPU time. The listing of a selection of these results, $3N_P N_Q(=693)$ triplet entries occupies 17 printed pages in this reference.

It turned out to be extremely difficult to draw overall conclusions from this data. Individual results were useful in locating specific errors or defects in particular routines or classes of routines.

But to find a basis for using this daunting mass of evidence to determine the best routine proved to be an almost impossible task. Kahaner made a brave attempt and selected three routines for the Los Alamos library.

At the moment, as readers will know, the search for appropriate testing techniques continues. A first step in this search was clearly to examine critically the battery testing technique to see what if anything was wrong with it and whether it could be improved. It was in the course of this search that the significance of the nature of the performance profile was first appreciated.

One of the technical difficulties in interpreting the results of Battery testing is attributable to the jagged nature of the performance profile. For example one could use routine A and routine B to integrate a particular function, perhaps one having a peak of a specified nature at a specified point. One might find that routine A was both more economic and more accurate than routine B. Then one could change the problem slightly by moving the peak a short distance and repeating the experiment. One could easily find that for this apparently similar problem, routine B was better than routine A. The reason in the end was attributed to the jagged nature of the respective performance profiles. That is, a minor change in the problem leads to a different conclusion altogether. If on the other hand the performance profiles had been smooth, minor changes in the problem would have led to only minor changes in performance.

A situation of this type embedded in a battery type testing procedure is unsatisfactory for several reasons. If for example some point system is used, in which one routine is credited with more points than another it it is better than the other on an individual problem, then the overall result depends on the lucky (for routine A) chance that one happened to choose the first problem and not the second when either could have been chosen. It could be hoped that this effect would even itself out when many different problems are involved but as far as I know there have been no experiments carried out to see to what extent this actually happens. But, much more significant is that one cannot extrapolate from a result of this type. The results of testing should if at all possible be presented in a form in which extrapolation is possible. A user should be able to expect that if his problem is reasonably close to one of the test problems, his results will be reasonably close to the test problem results.

Certainly, testing routines by examining individual problems is an important part of program development and will continue. But my belief is that the value of this sort of testing is limited. Large scale software evaluation projects will have to rely principally on the more sophisticated and expensive testing techniques which are designed with routines having jagged performance profiles in mind.

The construction of such techniques is quite complicated. At the present time, a technique for evaluation of Automatic Quadrature routines exists (see Lyness and Kaganove 1977) and one for unconstrained minimization down a curved valley (Lyness and Greenwell 1977, Lyness 1979). These are quite different from each other in detail; what they have in common is that they deal with software which has jagged performance profiles (as well as software having smooth performance profiles) and the numerical results are in a form in which a user can safely extrapolate to a similar problem.

5. A Software Interface Problem

The widespread use of items of numerical software has uncovered problems which can arise when one item is used in conjunction with another. I illustrate by an example. One wishes to find the maximum of $F(\lambda)$ when $F(\lambda)$ is defined by

$$(5.1) \qquad F(\lambda) = \int_a^b f(x,\lambda)dx = I_x f(x,\lambda)$$

and $f(x,\lambda)$ is an analytic function of both x and λ. Thus one could use an optimization routine whose objective function is calculated using a quadrature routine to evaluate the integral. If the quadrature routine actually produced exact results there would be no problem. However, the quadrature routine produces an approximation $Q_x f(x,\lambda)$ and so the optimization routine in fact attempts to find the maximum of

$$(5.2) \qquad \tilde{F}(\lambda) = Q_x f(x,\lambda)$$

which is a function close to, but not identical with $F(\lambda)$.

Most optimization routines work well when the objective function is smooth, but are slow and even unreliable when it is a rough or discontinuous function. Since the objective function is $\tilde{F}(\lambda)$ and not $F(\lambda)$ we should make sure, if possible, that $\tilde{F}(\lambda)$ is a smooth function. This depends on the quadrature routine. To see this one expresses $\tilde{F}(\lambda)$ in the form

$$(5.3) \qquad \begin{aligned} \tilde{F}(\lambda) &= I_x f(x,\lambda) + (Q_x f(x,\lambda) - I_x f(x,\lambda)) \\ &= I_x f(x,\lambda) + E(\lambda) \end{aligned}$$

where $E(\lambda)$ is simply the performance profile (2.5). Since $f(x,\lambda)$ is a smooth function of λ, so is its integral. But as we have seen, $E(\lambda)$ may be smooth or jagged depending on the quadrature routine used.

Thus, if we use an automatic quadrature routine for the integration, which produces a jagged performance profile, we set the optimization routine a more difficult task, that of optimizing a jagged objective function. This leads to an unnecessarily long and unreliable calculation.

A user who is unaware of the significance of the nature of the performance profile may easily wrongly diagnose the trouble to be that the optimization routine is inefficient. It must be emphasized that this trouble would occur when both routines are highly efficient, and working perfectly. It is termed an <u>interface</u> problem. And once the user understands what is going on, it is extremely easy to put things right.

There are several other aspects to this particular interface problem. Some of these are discussed in Lyness 1977 and in Lyness 1981. But the key result here is simply qualitative. It is important not to use an inner routine which has a jagged performance profile, but to replace it if possible by one which has a smooth performance profile. Not because individual quadrature results are better or worse, but because the jagged performance profile in the inner routine affects the working of another interrelated part of the overall program.

A similar effect occurs when using product one-dimensional quadrature to evaluate multiple integrals. In Lyness 1981 a detailed description of this particular effect is given.

The possible magnitude in terms of computer time is illustrated by an extreme example quoted in Lyness 1981. At Stanford one user found recently that a triple integration problem required three hours CPU time using product integration with three automatic quadrature routines. When the two inner routines were replaced by fixed abscissa rule evaluation routines, only one minute of CPU time was required to achieve the same accuracy.

6. Concluding Remarks

In this paper, I have done little more than to introduce the term performance profile, and to indicate how its nature (smooth or jagged) can affect numerical processes which involve the software. In this short survey it may appear to the reader that routines with jagged profiles are worse than those with smooth profiles. I believe that that is not a useful way to think about this problem.

The overall situation is that a particular item of numerical software has many properties. Important among these are accuracy, reliability, cost and so on. One of its properties is the nature of its performance profile. In any particular problem in which this software is applied, some of these properties may be important and others unimportant. I have mentioned several situations in which the nature of the performance profile is important. There are obviously other problems in which this is not important at all.

References

Kahaner, D. K. (1971). Comparison of Numerical Quadrature Formulas, Mathematical Software, pp. 229-259, J. R. Rice, Ed., Academic Press.

Lyness, J. N. and Kaganove, J. J. (1976). Comments on the Nature of Automatic Quadrature Routines, ACM Trans. on Math. Soft. 2, pp. 65-86.

Lyness, J. N. and Kaganove, J.J. (1977). A Technique for Comparing Automatic Quadrature Routines, Comp. Journ. 20, pp. 170-177.

Lyness, J. N. (1977). An Interface Problem in Numerical Software, Proc. Sixth Manitoba Conference on Num. Math. and Comp. 1976, pp. 251-263.

Lyness, J. N. and Greenwell, C. (1977). A Pilot Scheme for Minimization Software, ANL-AMD TM-323, Argonne National Laboratory, Argonne, IL 60439.

Lyness, J. N. (1979i). Performance Profiles and Software Evaluation: Chapter in Proceedings of W.G. 2.5 Working Conference on "Performance Evaluation of Numerical Software," ed. L. D. Fosdick, North Holland Pub. Co. (1979), pp. 51-58. Also available in ANL-AMD TM-343.

Lyness, J. N. (1979ii). A Bench Mark Experiment for Minimization Algorithms, Math. Comput., 33, pp. 249-264. Also available in ANL-AMD TM-323.

Lyness, J. N. (1979iii). The Affine Scale Invariance of Minimization Algorithms, Math. Comput., 33, pp. 265-287.

Lyness, J. N. (1981). When Not to Use an Automatic Quadrature Routine (submitted to SIAM Review).

The Role of Computer Centers in the Field of Mathematical Software

P. C. Messina*

Applied Mathematics Division
Argonne National Laboratory
Argonne, Illinois 60439

A. Introduction

I have been involved with mathematical software for approximately fourteen years. I set up scientific subroutine libraries for an industrial company and a university, was responsible for mathematical software libraries at Argonne National Laboratory for five years, and was manager of the Mathematical Software Project of the IBM users organization, SHARE, for two years. A long time ago I even wrote some routines. Currently, I help coordinate several mathematical-software related projects at Argonne National Laboratory. My remarks are based on these experiences, various publications on the topic of mathematical software libraries, and conversations with many others who have had similar responsibilities. At the outset I would especially like to acknowledge the contributions of Dr. Jesse Y. Wang, with whom I worked for several years and who is currently responsible for the support of mathematical subroutine libraries at Argonne.

I am still amazed how difficult it is to convince users to take advantage of libraries of mathematical software which, nowadays, are of very high quality. Although the programs are much better written than before and their documentation is generally quite good, often the routines are not used [1]. The computer center can have a major influence on the utilization of library routines. It is

* This work was supported by the Applied Mathematical Sciences Research Program (KC-04-02) of the Office of Energy Research of the U.S. Department of Energy under Contract W-31-109-Eng-38.

responsible for selecting, installing, publicizing, and helping with the use of the libraries. Its staff has first-hand knowledge of user needs and problems with the software. Thus, computer centers can play an important role in the field of mathematical software.

B. Justification for Centrally Supported Mathematical Software Libraries

Many computer centers do little with mathematical software; at best, they make available a commercially produced library or a collection of locally written routines on an "as is" basis. The lack of support is sometimes justified by reasoning that computer centers cannot be expected to provide libraries of software in all fields. On the other hand, mathematical software has features that qualify it for special treatment:

- Mathematical computations are essential in an extremely large number of fields.

- It is possible to write subprograms that are general enough to require no modification to be used in many applications.

- Promoting the use of libraries tends to be for the common good; that is, duplication of programming effort is reduced, as is the use of machine resources to support it. The numerical results obtained will, in general, be more accurate and trustworthy. Routines written by experts can also be more efficient than those written by casual users, especially when they are designed for computers with sophisticated architectures such as CRAY-1, CDC 7600, and IBM 370/195.

- For a large computer center, mathematical libraries can be supported with modest expense. When a computer center has many users, the cost of supporting mathematical software libraries is typically distributed through a slight increase in the machine usage rates. If good libraries are available and well supported, there should be enough users (big and small) who actually say they want such support and are willing to pay the modest additional cost. At Argonne the cost is about two per cent of the total budget for operating the central computer facility.

- Successful use of library routines may encourage modular programming. With modest effort, one can create programs quickly by calling routines from existing libraries. A favorable experience may seduce the user into better program design.

These arguments tend to apply best to rather large computer centers with a varied user community. Small computer centers with a more specialized community may find the situation different. Also, they clearly will not be able to spread the cost of the mathematical libraries over as big a user base, and so, the cost will be more noticeable and perhaps too large for their situation.

C. Communications Between Users and Suppliers

Authors of mathematical software use different criteria from users to judge software products. Computer centers can help each group become more aware of the other's set of values. The differences are characterized below.

1. The Authors' View of What is Important

What I am about to say will no doubt include gross generalizations and be unfair to many authors of mathematical software. My remarks are based on observations made at conferences on mathematical software, conversations with authors, and publications on the subject. The following qualities are highlighted:

- Accuracy of results

- Newness of algorithms used

- Speed at least on some problems. Unfortunately, measurements of execution speed are often carried out poorly, and differences of a few percent are given more importance than is warranted [2]. Relatively slight changes in versions of compilers can reverse the results as to which routine is faster [3]. The same can happen when different computers are used [4]. A program may be faster than another when both are run on an IBM machine, yet slower when they are run on a CDC computer. On some machines, even small changes in the Fortran source can result in very large changes in execution speed. For example, on extreme cases on the IBM 370/195 at Argonne we have seen changes in speed of a factor of 10, as a result of changes in the Fortran source and not in the algorithm. (The 370/195 has multiple, segmented floating-point arithmetic units, an instruction pipeline, and a high-speed buffer memory; exploitation of these features can lead to dramatic reduction in run times.)

- Robustness. I am, of course, referring to "robustness" as defined by Cody [5]; that is, the ability of a program to notice bad input and to notify the user of the situation, instead of giving wrong results. While robustness is an important property of mathematical software, I believe it should not be emphasized to users: doing so implies that users are careless or stupid and forces authors of mathematical software to work hard to build in robustness.

- Good documentation of the algorithm. This documentation often appears in the form of a journal article published in the open literature. Typically it is well written for other numerical analysts to read.

Two other qualities, ease and convenience of use, have also been given a lot of attention and emphasis in the design of software collections [4], [5].

2. The Users' View of What is Important

- Algorithms that solve the users' problems. Typically it does not matter to users whether the algorithms are up-to-date or not. It is more important that the input necessary for their operation can be supplied by the user and that the algorithm, in general, solves the problem.

- Speed of execution. The users, too, are very preoccupied with speed; and, unfortunately, the same comments made earlier apply to the quality of their methods for measuring speed. For some reason that is not completely clear to me, subprograms from libraries have the reputation of being slow. I have encountered users who will not even consider using a library subprogram because they "know" that it will be too slow. When asked, these users admit that they have not bothered to measure the speed of the library program because they "knew" ahead of time that it would be too slow. Large organizations are considered incapable of writing efficient code. While the users also value speed, accuracy, and good documentation, their emphasis differs somewhat from the authors of mathematical software. The following qualities are highlighted:

- Ease of use

- Documentation that they can understand

- Modest accuracy requirements. Frequently the user will be able to get by with less accuracy and enjoy the resulting higher speed. Of course, I feel that if it is possible in an algorithm to achieve high accuracy, one should certainly do so. But, on the other hand, if there are cases where it is possible to give the user the option of trading execution time for accuracy, one should attempt to do so. Frequently the user will be able to get by with less accuracy and enjoy the higher speed that results.

To illustrate some of the points that I have made above, consider an experience which I had some years ago. In 1967 I was hired for the summer by an industrial company to help a chemist use the fast Fourier transform algorithm for some computations. He had been carrying out the computations by numerical integration of the Fourier integrals involved. The central computer facility for his company had a scientific subroutine library that contained a fast Fourier transform routine, so it seemed like a trivial task at first. However, I needed several weeks of nearly full-time work to understand how to use the routine, and I ended up modifying it and writing ten pages of new documentation. When I was finishing the documentation, I happened to talk about my project to some users and learned that two of them had tried in vain to learn how to use that same subprogram, had finally given up, and were still using older and much slower techniques. This is a special case because the FFT gives an unusually large advantage in speed of execution over other methods and is also, in general, more accurate. The algorithm had just been rediscovered in those days and was very well known, and yet, even with all those advantages, several interested users had given up the idea of using it. I was hired for the express purpose of making that algorithm available, so I could dedicate the time necessary to learn how to use it. Most people simply do not have so much time to dedicate to such a task.

D. Other Factors That Influence the Use of Mathematical Software Libraries

The previous section stated that authors and users of mathematical software emphasize different things and this can result in less utilization of the libraries than one would expect. There are additional factors that influence how much libraries are used, not all of which can be controlled by the authors of mathematical software.

1. Accessibility. Under this topic I refer to how easy it is to determine the contents of the library; obtain documentation for the library routines; get detailed information on what problems can and cannot be solved by a given subprogram; gain access to on-line executable versions of the library; obtain source for the program; and obtain versions of the program for all the computers of interest to the user. (On this last point I might note that it is becoming common to develop a program on one computer and then run it on several different computers, all at the same site. If the libraries available on each computer are different, then their use is severely restricted.)

2. Completeness of Coverage. It is important that libraries contain routines that address many different kinds of computations. One reason is that if a library has a limited scope, then users will not be inclined to consult even the table of contents for the library when they need a routine for a given computation. The broader the coverage, the more likely it is that they will acquire the habit of consulting it.

3. Stability. I have in mind several different kinds of stability:

- The routines are never changed without prior notice to users.

- Libraries are made available quickly for new computers. In this way there is no disruption in the availability of libraries as the computer center upgrades its equipment.

- When library routines are modified, users can still access old versions of the routines for a time. This feature is important for many people who have to run benchmark programs periodically to control that they are still getting the same answers. It is essential to these users to be able to determine what is causing a difference in results, whether it is a change that they made or a change in a library routine or some other part of the system.

- One should try to have no bugs in the library routines to begin with; if bugs are found, one should fix them properly as soon as possible. If there are a few instances of bad fixes applied to library routines, users will feel there is an unstable situation and avoid using the library in the future.

- One should convey the firm impression that there will be continued support of mathematical software.

4. Expert Consulting. The computer center should provide people who will act as consultants on the mathematical library.

5. User Opinion. There are several ways to give users a voice in how libraries are supported. At the very least, one can request user opinion on proposed important changes to the overall structure of the libraries, perhaps, or breadth of coverage. Generally, there are a few users who are willing to take the time to discuss proposed changes and give their opinion; they often have useful suggestions to make. In addition, one can make available to users new editions of libraries on an experimental basis to try out as they see fit before they become production versions. If the users feel that there are significant enough problems to postpone the granting of production status to a new library edition, then one should take their advice.

F. Role of the Mathematical Software Librarian

The mathematical software librarian can perform a crucial role in helping both users of mathematical software and authors. What must this person do?

- Choose and install libraries and make their documentation available. The high quality of commercial mathematical software libraries nowadays makes this task relatively straightforward.

- Consult, both on ways to use the routines and on the mathematical techniques used. The choice of algorithms is often given little thought by the user. The librarian should, in general, try to find out what the user really needs to compute, instead of answering an immediate question on the detailed usage of a particular routine.

- Investigate bug reports. Most apparent bugs in mathematical software are actually problems with the usage. Mismatched arguments are a common example; however, occasionally there are real bugs. In this case, the librarian should inform the author of the routine immediately.

- Notify authors of mathematical software of bad human engineering features in their software or documentation; and also inform the authors on needs for new methods or routines. In my experience, authors are very receptive to this kind of information since they frequently have little personal contact with users of their software.

G. Conclusions

Mathematical software libraries have the potential of improving the reliability and reducing the development cost of scientific programs. The ability of mathematical software libraries to achieve this potential is reduced by a number of factors. Some can be addressed by the producers of mathematical software; however, there are factors that can be addressed only by the computer center. A mathematical software librarian can help both users and producers of mathematical software.

The quality of commercial libraries has increased considerably in the past decade and has reduced to a reasonable level the cost of providing mathematical software at large computer centers. The future is not as bright: the availability of inexpensive yet powerful computers is resulting in the establishment of many new, relatively small computing centers with an even wider variety of different computers and operating systems than in the past. This phenomenon makes it increasingly difficult to provide access to libraries of mathematical software to many users.

References

[1] Einarsson, B., Erisman, A. M., Hitotumatu, S., Shampine, L. F., Wilkinson, J. H., and Yanenko, N. N., "The Use of Mathematical Software Outside the Mathematical Software Community: A Panel Discussion," Performance Evaluation of Numerical Software, L. D. Fosdick, (ed.), North-Holland Publishing Co., New York, 1979, pp. 273-283.

[2] Crowder, H., Dembo, R. S., and Mulvey, J. M., "On Reporting Computational Experiments with Mathematical Software," ACM Transactions on Mathematical Software, Vol. 5, No. 2, June 1979, pp. 193-203.

[3] Parlett, B. N. and Wang, Y., "The Influence of the Compiler on the Cost of Mathematical Software," ACM Transactions on Mathematical Software, Vol. 1, No. 1, March 1975, pp. 35-46.

[4] Ford, B., Hodgson, G. S., and Sayers, D. K., "Evaluation of Numerical Software Intended for Many Machines - Is it Possible?", Performance Evaluation of Numerical Software, L. D. Fosdick (ed.), North-Holland Publishing Co., New York, 1979, pp. 317-329.

[5] Cody, W. J., "The Construction of Numerical Subroutine Libraries," SIAM Review, Vol 16, No. 1, January, 1974, pp. 36-46.

[6] George, A., and Liu, J. W. H., "The Design of a User Interface for a Sparse Matrix Package," ACM Transactions on Mathematical Software, Vol. 5, No. 2, June 1979, pp. 139-162.

[7] Gill, P. E., Murray, W., Picken, W. M., and Wright, M. H., "The Design and Structure of a Fortran Program Library for Optimization," ACM Transactions on Mathematical Software, Vol. 5, No. 3, 1979, pp 259-283.

Guidelines for Managing Mathematical Software Libraries at Computer Centers

P. C. Messina*

Applied Mathematics Division
Argonne National Laboratory
Argonne, Illinois 60439

I. PROCEDURES

Background

The guidelines I have compiled for managing mathematical software libraries are, of course, based largely on my own experiences, supplemented by experiences I have read about. While I present these guidelines in the hope that they will be helpful to others who are offering mathematical software libraries to a general user community, I strongly believe that each environment tends to be different and that techniques which have worked effectively at some installations may well need to be modified to be much good at others.

In this paper I will refer most heavily to the mathematical software libraries at Argonne National Laboratory. The Central Computing Facility at Argonne has had several large-scale IBM computers since the mid-sixties. This particular type and size of equipment have affected some of our decisions.

---

* This work was supported by the Applied Mathematical Sciences Research Program (KC-04-02) of the Office of Energy Research of the U.S. Department of Energy under Contract W-31-109-Eng-38.

Choosing and Organizing the Content

At Argonne we have found it useful to segregate different collections of mathematical software into different categories, which we make known to the users. This has several operational advantages. The different categories of mathematical software receive somewhat different levels of support. As will become more obvious later, it is fairly difficult, if not impossible, to prepare and keep up to date a comprehensive library that contains items from many sources and make it look to the user as if it is all one integrated library. At Argonne we distinguish five categories of routines.

1. Locally supported routines. This library is comprised almost exclusively of routines that were written locally. In addition, general collections like EISPACK [1], FUNPACK [2], LINPACK [3], and MINPACK [4] -- which were written at Argonne -- are included in the library. These are locally supported and receive the best support level.

2. New routines that are locally written, but not yet fully tested. We like to make these available to users for testing prior to declaring that they are ready for production use.

3. Obsolescent routines. Periodically we announce that certain routines are no longer the best available. These are kept for several months before we withdraw their availability completely.

4. Routines obtained from elsewhere, but not from commercial libraries. These routines are of high quality or we would not have gone to the trouble of getting them. However, the suppliers make no commitment to maintain them. At Argonne we put such routines in a library we call SYS2.AMDLIB. An example is the package for solving elliptic partial differential equations, which we obtained from the National Center for Atmospheric Research [5]. This is a good package and has fairly good documentation; however, because the routines were not written at Argonne, we do not have the expertise to support them in the same way as locally produced software.

5. Commercial libraries. We lease the IMSL and NAG libraries. Bug fixes and new editions are provided by their vendors.

We maintain four separate disk files for the above five categories: SYS1.AMDLIB, which contains the routines in category 1; SYS2.AMDLIB, which contains the routines in categories 2, 3, and 4; the IMSL library, from category 5; and the NAG library (also from category 5). (We split these two commercial libraries to make it easier to install new editions.) We keep on-line executable versions of all four disk files. On IBM systems, at least, it is very convenient to access such libraries, as opposed to recompiling the routines each time. SYS1.AMDLIB is automatically searched every time a FORTRAN or PL/I program is link-edited to prepare it for execution. This feature enables users to CALL routines from this library with no changes in their job control language statements. In addition, at Argonne we have provided special "catalogued procedures" that greatly simplify the job control language needed to search several libraries in any desired order. We believe the extra effort is worth it to facilitate access to the libraries.

For the non-commercial libraries, we make source readily available to users upon request from a librarian; the librarian, by determining the user's specific needs and satisfying them properly, avoids many user problems. Consider the following example. If a casual user were to obtain the source for a FUNPACK routine to run on another computer, say a CDC computer, he will discover that the conversion is difficult if accuracy is to be retained. The librarian, however, could properly steer the user to the appropriate Control Data version of FUNPACK.

For the commercial libraries, we help users deal with IMSL or NAG to obtain permission for the source, if appropriate. Typically, if the routine is incorporated in an application program, it is possible to get permission from IMSL to redistribute the source, even to other installations that do not have licenses for the library.

We, too, have discovered that free software is not always a bargain. We are very cautious about accepting and installing collections of routines from elsewhere which cost nothing to purchase or lease, but may require a great deal of work to support in any fashion. We prefer not to make available to users mathematical routines with which we are not familiar and which may produce bad results. If users want to acquire such software themselves, they are free to do so; but we do not want to encourage getting a lot of things for which the reliability and accuracy are not known.

Adding New Routines

In trying to decide whether to add new routines or libraries or not, we consult experts as much as possible. We are fortunate to have a fairly wide range of expertise locally at Argonne, so we can get good advice in many areas of computation. On occasion, we will seek out other people's advice as well.

The details of the process used to add new routines vary from category to category.

1. Locally Produced Routines.

For locally produced routines, even if they are part of "packs," we require the following materials:

- Documentation in machine-readable form -- more about this topic later.

- Source listing.

- Source deck, with in-line documentation.

- Compilation listing.

- Object deck.

- Test deck. We ask that this test be as exhaustive as possible. Of course, there are no really comprehensive tests, and sometimes they are quite minimal; however, the instructions to the author state that the test should be as exhaustive as possible.

- Test case results. Of course, we want to be able to see what results the author got with the test case. If there are more than 50 pages of output, we request that these results be supplied on micro-fiche instead of on paper.

- Demonstration program. This program is intended to illustrate a typical use of the routines involved.

• Output listing of demonstration program. We request a listing of the output of the demonstration program so that we can compare our results with those obtained by the author.

By requesting these materials, we have found that we generally receive rather high-quality routines which are not too difficult to maintain. Apparently, people who take the time to supply all the materials we request do a fairly good job of writing the routines.

2. External, non-commercial sources

For routines from external, non-commercial sources, such as the NCAR package for elliptic partial differential equations, we cannot expect authors to have already produced the kinds of materials we want and the format that we normally use for locally written routines. Consequently, we use our judgment as to when the materials available are good enough for us to accept the package. The level of our standards also varies according to our perception of how useful the package would be for our user community: The more useful it is likely to be, the more work we are willing to undertake to make it available properly. We hope good documentation is already available because that is the most time-consuming thing to produce. In a few cases the documentation has been good, but not in machine-readable form, and we have taken the time to put it in machine-readable form. A test program should also be available from the author fairly easily. The remaining materials, such as the compilation listing or an object deck, we can fairly easily produce ourselves.

3. Commercial Libraries

While the commercial libraries now available do provide good materials and have high-quality routines, in general, we would hesitate to get another one (in addition to IMSL and NAG), not because of the lease cost (which we feel is modest), but because of the effort required to support even a well-packaged commercial library. For example, these libraries tend to have a major new version every 12 to 18 months, and it requires quite a bit of work for the staff to familiarize themselves with the changes, provide some local documentation, and process each new version.

Advertising New Routines

Once new routines or libraries are selected and all the materials are gathered and processed, the final task is advertising their availability to the user community and making documentation available. For commercial libraries, for example, manuals have to be ordered; and it may take weeks or even months for them to arrive.

Updating Existing Routines

The procedure for updating locally written routines involves four steps:

• Whenever a change is made to a routine we always notify the users, no matter how insignificant the change. We believe it is important for users to know what is going on.

• If the usage of the routine changes very much, we try to issue the changed routine with a new name.

• We update the documentation of the routine, note the reason for each change and the date of the change, and update the program statistics, such as the number of source statements of the program and the size of the compiled program.

• We rerun any tests that are available for that routine, as well as the demonstration program. If the change was to fix a bug, we try to improve the test so that future tests might have some chance of catching that bug before the routine is released.

• When a serious error is found in a routine, say wrong results are produced for fairly typical usage, we immediately remove the routine from the on-line libraries. We still have to advise users of the problem, because some users may have incorporated the routine into their own programs. If possible, we indicate to users what circumstances trigger the bug.

Much the same procedure is followed when we update routines from outside sources. However, as an additional step, we try to notify the author of the routine about the changes that were made or the bugs that were found. the routine is from somewhere else, we will not usually be as familiar with it as with a locally written one. Therefore it may take us somewhat longer to find bugs; we may have to contact the author to request his or her help in fixing the routine. For anything at all serious, in fact, we do so. When the outside source is a commercial library, that library of course provides a new edition or fixes to existing routines: We merely have to apply the changes and notify users of what has happened. Again, we do not change the contents of the libraries without first notifying the users.

Removing Routines from the Library

Routines are generally removed because better ones have become available. We announce the planned deletion of a routine to users three months in advance, and we stop distributing its documentation. Occasionally users insist on continuing to use the obsolete routine. In those cases we will give them a source deck, if it is not from a commercial library that prevents us from doing so. We feel that we must give advice to users about what is a good routine and what is a bad one; but if users insist on using a bad one, they have the right to do so. At Argonne, if the routine to be removed was in SYS1.AMDLIB, which is automatically searched, one month after the initial notice of removal the routine will be moved from the automatically searched SYS1 library to the SYS2 library. (Recall that the SYS2.AMDLIB library is used at Argonne for obsolescent routines as well as other things.) This technique keeps the routine available to users for some period of time, but, by not having the routine in an automatically searched library, reinforces the notices that the routine will soon be deleted. Finally, we remove the routine from the SYS2 library after about three months.

Installing New Editions of Commercial Libraries

We put new editions of commercial libraries in a separate disk file and, after some minimal testing, make them available for user tests. The user test period is never less than a month and may last several months.

Next we advertise the new capabilities the library and the names of routines that are being deleted. Typically, the vendor of the commercial library provides some good draft articles for this purpose.

Once the new edition has been adequately tested, it becomes the new "production" version. That is, it is moved to the appropriately named disk file; the old production version is moved to another on-line disk file so that easy access to it is retained for a few more weeks. New documentation, as needed, is made available to users.

Ensuring Uniform Documentation

Documentation guidelines depend on whether a routine is produced locally or externally.

1. Locally Written Routines

For maintenance purposes we believe that internal documentation is very helpful; therefore, we ask authors of subroutines to try to make their source code readable. We suggest that they use mnemonic variable names; that they use indentation to show the structure of control flow; and that they include extensive comments, for example, at each branch point, at the beginning of each block, and at each call or function reference, explaining the intent of the code. The general recommendation is that the author try to make the code self-explanatory.

Our external documentation requirements have evolved over the years to the format described in Appendix A. The documentation is intended to be used both by the librarian and the prospective user of the subroutine. Our general format lists a wide variety of information; any one routine will not include all those items in the documentation. We deliberately try to be flexible in our documentation requirements, because too rigid a structure will make it inappropriate for a certain class of routines. We do require the documentation to be machine readable. Appendix B lists our conventions for representing mathematical notation. Appendix C contains the documentation for a routine in the Argonne library.

2. Packages and Commercial Libraries

Uniform documentation is hopeless to achieve when externally produced libraries or packages are involved. External suppliers each have their own standards, production methods, etc., and it would be far too much work for us to convert their documentation to a uniform format throughout. Therefore, we resign ourselves to living with documentation that is not uniform. On the other hand, the documentation supplied with commercial libraries nowadays tends to be quite good.

3. Guide to all Mathematical Software

So far I have been concentrating on documentation for each individual routine. An extremely useful document, however, is one that describes what areas of computation are covered by the various libraries available and gives advice on which routines to choose when more than one routine can do a particular computation. This document should also give information on the areas of numerical analysis for which local expertise and in-depth consulting are available. Few installations that I know of rely on only one commercial library to satisfy all of their mathematical software needs. Therefore, almost everyone needs to produce such a document to guide the user in selecting routines from the various collections of mathematical software.

4. Index of Holdings

Another locally produced document that can be quite useful is a consolidated index of holdings arranged by subject area. It helps users determine what areas are covered and how thoroughly. It should be supplemented by a keyword in context index. The latter contains one-line descriptions of each routine; each keyword in the description is used to produce an index entry with a reference to the routine. So, for example, under the word "minimization" would appear all the one-line descriptions that contained that word, along with the routine names.

II. SUPPORT POLICIES

Background

This section continues the discussion of guidelines for managing mathematical software libraries at computer centers. In the previous section, I discussed procedures for choosing, installing, and documenting different collections of mathematical software. Here, I focus on the issues involved in supporting mathematical software libraries.

Debugging, Consulting, and Education

To ensure that mathematical software libraries are used effectively, the computer center must respond quickly to user problems and should help the users in selecting and utilizing library routines.

1. Debugging

It is extremely important that the computer center react quickly to reports of bugs, even though most such reports will be due to user error (say the wrong number of arguments in the calling sequence). One should treat each error report as if it may have uncovered a mistake in a library routine and determine quickly whether that is the case. If, indeed, a bug must be fixed, a "quick fix" is desirable, but the users should understand that a final solution may take a few days, or even a few weeks.

2. Consulting

A consulting service is also important. There are two things I wish to emphasize in providing such a service:

First, the people who do the consulting should be willing, as well as able, to help users choose and use library routines. Choosing routines for a given problem is often difficult; and having someone available who believes it is his or her responsibility to help make the choice can result in more and better usage of the library. If the right choice is made initially, and some help is provided on how to use a routine, then fewer spurious bug reports will occur later.

Second, it is occasionally possible to help users, not just with choosing routines, but also with formulating the mathematical model or selecting the computational method to be used to solve it. When the expertise is available and the user is receptive, then such help should be provided. After all, the real goal is to have a user obtain the answers he needs, not merely to help him use some software.

A useful byproduct of offering a consulting service is that it provides information on user needs for routines which are not yet available. In the course of helping users with their problems, one will discover gaps in the library. Without such contacts one may never know of the need for certain routines.

3. Education

Two levels of education can be offered profitably. One would be brief (say one- or two-hour) seminars given periodically, that describe the local libraries, the level of support available for them, the on-line files they reside in, and the procedures for accessing their contents and obtaining documentation. The second level of education would be workshops on some specific computational area, such as optimization. These workshops would last several hours, spread over two or three days, and would review the mathematics and numerical analysis involved, as well as describe in considerable detail the software that is available locally to solve such problems. Bell Laboratories and Stanford University have both had good experiences with such workshops for several years.

Offering these workshops establishes human contacts, which are invaluable. It has been my experience that users who have attended a course I have taught are much more likely to call me later when they need help with something--even something not directly related to the course they attended. In addition, the workshop plants in users' minds the idea that there are people available who are knowledgeable about numerical computation and who are willing to take the time to help others do it. Indeed, even if people do not attend the workshop but only read the announcements, they may get a more favorable impression of the support of numerical libraries available at their center.

Staffing Considerations

Although earlier I devoted many words to outlining procedures for supporting mathematical software libraries, I believe that those procedures are secondary to the quality of the people available to carry them out. The choice of personnel is the most important single factor in providing good support to mathematical software.

The staff must be familiar with a broad range of computational methods, as well as software to carry out those methods. They must also be familiar with the overall computing environment at their installation, in order to give sound advice about the efficiency of certain strategies, and to debug routines effectively without being misled by quirks of the rest of the computing environment. In addition, those who maintain libraries and process new routines must be extremely careful and thorough workers: quality control is important to secure and retain user confidence in the libraries; not every good mathematician has the tendency to be this careful at these tasks. Finally, the staff must be able to communicate well with users.

It is impossible for the computer center to have experts in every area of interest to its user community. However, there frequently are people in the institution, as a whole, who are expert in some of the areas and who may be willing to provide consulting advice, especially if the computer center undertakes to screen out the more trivial questions. Sometimes people from other institutions are also receptive to being asked for advice on more challenging problems in their field of expertise. For example, university professors are frequently hungry for contact with real applications.

• The number of large "production" programs that use library routines. Clearly, the ideal type of person for providing mathematical software support must possess a considerable range of rare skills. Fortunately, even large computer centers can get by with one (or occasionally even slightly less than one) such full-time person.

Attitudes Towards Users

One should not assume that library routines are always best for a user's problem. Keeping an open mind not only will eventually attract more users to a library, but should result in better advice. For example, a few years ago at Argonne a user asked for help on tuning his code, that is, making it execute faster. By running his program through an execution analyzer, we discovered that approximately half of the execution time for that program was spent in evaluating the square roots of numbers (using, of course, the library-supplied square root routine). That routine was written in Assembler Language rather carefully, so that it was efficient; however, it was by necessity a general-purpose routine. For this particular user, after the first call to the routine all subsequent calls were with arguments that were very near in value to the original argument. Therefore, it was possible to write a specialized square-root routine (and put it in line, incidentally, to avoid the overhead of subroutine calls) that exploited this knowledge about the range of arguments. This approach saved nearly 40 percent of the total run time. Therefore, even in seemingly trivial library routines there can be room for improvement for a particular user problem.

Modification of library routines is occasionally desirable. In those cases where it appears worthwhile to enhance or specialize a routine, one should try to enlist the original author's aid. The author tends to be receptive to such modifications if they will better handle a class of user problems. It should take less effort to modify a high-quality routine rather than write a brand new one, to exploit some particular feature of a user problem.

A librarian should try to determine (as tactfully as possible) whether the user has formulated the computation in an undesirable way rather than just answering the question posed by the user. The typical example here is a user request for a matrix inversion routine; one asks whether the inverse of the matrix is really needed for its own sake or whether the user wants to solve a system of linear equations. Naturally the latter situation is the most common, and there are better ways to solve linear systems than by inverting the matrix of coefficients. In any case, the idea is to help the user as much as possible with his overall problem, not with his immediate problem.

Characteristics of Available Mathematical Software

A few comments on features of available mathematical software illustrate a librarian's concerns.

1. Name Problem

The variable names used for arguments and routines are sometimes not user-oriented. An example, I regret to say, is found in the LINPACK collection in whose development Argonne played a major role. Several routines in LINPACK have an argument called "P" of type integer. Since in Fortran a variable named "P" is assumed to be REAL unless it is explicitly declared to be INTEGER, the documentation prominently states that "P" is type INTEGER. Nevertheless users periodically forget to declare "P" to be an integer and get mysterious execution errors. The variable is used to represent the order of the input matrix; therefore, there is no particular mnemonic value to using "P" except for some numerical analysts. However, the typical user of these routines is certainly not a numerical analyst.

Many other libraries and collections of mathematical software have chosen strange and unpronounceable collections of characters as names for their routines. It would be nice if subprograms could be named in some mnemonic fashion to make it easier to remember which ones to use. Unfortunately, this is not possible. There are so many routines available nowadays at a typical center that it is impossible to use mnemonic names any more without conflict. Therefore I am resigned to having to deal with strangely named routines for many years. Maybe some day in the future we will be able to use names of reasonable length and reduce the problem somewhat.

2. Portability

It is becoming more important that software either be portable or exist in versions for many different machines. Fairly often, users at Argonne need to send a program they have developed to someone at another installation. If the mathematical routines that are used were written portably, then it will be easier to export the programs to some other environment. Moreover, with distributed computing becoming more prevalent, institutions need to have the same libraries installed on several different types of machines. Finally, when new machines are installed, libraries need to become available for the new machines quickly.

3. Availability of Source

Sometimes the source must be examined to ascertain exactly what method is used and how it is implemented. This need arises, for example, when the documentation is not detailed enough to determine the suitability of a given routine for a specific problem. Additionally, one must have the source in order to modify a routine to exploit features of the user problem. Finally, source is an invaluable aid for investigating bug reports.

4. Documentation

The cost and delivery time for commercial library documentation can be a problem. Such documentation is expensive, and we are therefore reluctant to keep large numbers of manuals available, especially since they become obsolete within a year or two. On the other hand, it takes weeks to months to obtain delivery of manuals from the time one decides to order them.

I hope vendors of mathematical software begin to provide machine-readable documentation for their libraries. This will enable us to customize the documentation; for example, one could add information on where the library resides locally and how to access it. In addition, one could more easily print subsets of the documentation on demand, instead of ordering additional copies of the entire library documentation. At Argonne we have taken this approach with machine-readable manuals for some other systems and find it quite useful.

One obstacle to machine-readable documentation has been the difficulty of providing aesthetically pleasing documentation for mathematics using the commonly available text formatting tools. However, two systems may considerably improve this situation. The UNIX system licensed by Western Electric includes software that provides a way to phototypeset mathematical material fairly easily and with modest hardware requirements. However, this software runs on only a few computers and is moderately expensive to buy. Soon the TEX system from Stanford should become available for a large variety of environments and may become a standard in the mathematical journal business. Currently, TEX is available for a moderate distribution cost--on the order of 30 U.S. dollars. If the goals for TEX are achieved, it could be used for mathematical software documentation.

Measuring the Effectiveness of the Service

So far I have spent a great deal of time talking about the desirability of supporting mathematical software libraries and consulting on their use; however, all my arguments have been qualitative rather than quantitative. Is it possible to measure the effectiveness of the service?

A. Information Desired

There is much information one might want to measure, if it were technically feasible to do so. A few that come to mind are:

- The number of distinct users of each routine, and their identity.

- The number of distinct users of the library as a whole.

- The number of executions of each routine in a given period of time, say every month.

- The amount of cpu time spent in library routines. This number could be used to estimate savings in cpu resulting from the use of supposedly more efficient library routines.

- The number of lines of source code from library routines that have been incorporated in user programs. This number could be used to estimate the effort saved in using a library routine rather than write a new subroutine. Studies of programmer productivity estimate that less than 10 lines of debugged and documented code are produced per person per day. Therefore, if one uses a 200-line subroutine from a library, which would have taken 20 days of effort to write, and if it takes 2 days of effort to learn to use it and incorporate it into one's program, then there is a net savings of 18 days of effort.

- The values of input arguments when routines are invoked by user programs. In some cases authors might use this information to produce specialized versions of the routines that were optimized to the range of argument values not frequently used in real applications.

• The number of large "production" programs that use library routines. Often in computer centers there are relatively few groups who develop and run rather large programs and consume a large percentage of the machine resources. Since they pay a large percentage of the bill, one might want to know to what extent these groups are taking advantage of mathematical software libraries.

• The number of times users consult the library holdings list for software to do their computations rather than just sit down and write their own new routine. Having on-line programs that users could browse through not only would encourage users to look at the available library holdings, but would make it possible to record the usage of such programs.

• Improvement of user end results as a consequence of using library routines. This type of information is, of course, very difficult to obtain. However, it would be the most useful data in a cost/benefit analysis for mathematical software libraries. Occasionally such information is available. For example, with the help of one of our staff, a program was written at Argonne's Chemical Engineering Division that models the absorption of tritium in the walls of a fusion reactor. This program used a variety of mathematical software routines:. the well-known Gear Ordinary Differential Equation package; several MINPACK 1, LINPACK and EISPACK routines; and VMCON, a least-squares fitting routine recently developed at Argonne. The program was able to obtain results that matched experimental data, whereas previous programs written elsewhere had not been able to do so. Since the mathematical model used was not different, the mathematical software deserves to get the credit for the improved results. I must admit that this is an isolated example; we don't know of many like it.

B. Measurement Techniques

Unfortunately, much of the data mentioned above cannot be measured on most systems. Here is a brief survey of what is possible:

1. Automatic monitoring of library usage

Operating systems have been modified to provide certain library usage information. Sandia Laboratories Albuquerque and Livermore, for example, modified the CDC SCOPE system [6]; and the Stanford Linear Accelerator Center made some modifications to IBM'S OS/MVT and VS2 operating systems [7]. Sandia's modifications provided information on each user who loaded a library routine from a mathematical software library, the routine name, the value of one argument on the first call to the routine, and the cpu time used on that run. The SLAC modifications recorded the name of routines loaded from system libraries into a user program and the identity of the user.

Unfortunately, neither of these modifications is widely available yet. The Sandia and SLAC modifications apply to obsolescent operating systems. Furthermore, as far as I know, they were never packaged and documented in such a way that they could be easily applied at another installation. It seems strange that although most modern operating systems rely heavily on software libraries to organize their work, there are no convenient measurement tools for determining which routines are used, how often, and by whom.

2. Monitoring of user inquiries and bug reports at the consulting desk

By keeping a log of user inquiries and bug reports and what routines were involved, one can get some information on routines of interest to the user community.

3. Recording the number of documents distributed (or scanned on-line) for each routine

This, again, is only partial information and is available only when separate documents exist for each routine.

Testimonials of good and bad experiences, although impossible to process statistically, can be useful. If users state that they find certain routines to be useful for their work, then that statement can certainly be used in trying to justify continued support of mathematical software libraries. Similarly, if there are reports of numerous bad experiences with routines, one at least knows that those routines are being used.

C. Data Interpretation

I have listed various types of information that I think it would be interesting to measure and noted the few, rather imperfect, ways that are currently available to measure some of this information. Now let us make some guesses about what the measurements would yield if we made them and offer comments on how to interpret them.

I suspect that if we monitored the usage of most mathematical software libraries, we would find out that most routines are used very seldom. Should those low usage counts then be used as an argument for deleting the routines from the libraries? Certainly not: Rather, I think one should expect small usage counts for most routines. It is after all, time-consuming to learn to use routines. Moreover, some computational methods are new to the average user and therefore will not be chosen very often. One example would be the use of spline-based packages for function approximation. Many current users were not introduced to splines when they received their formal education; being unfamiliar with the concept and the terminology, they will avoid such routines. Finally, the breadth of coverage of our libraries, especially the commercially available ones, makes low usage statistics for many of the routines likely at any one computer center. There are so many areas of computation covered that a given user community may not need all of them. Nevertheless, I believe that rarely-used routines should be in libraries, ready to be used. This is the archival function of a library, just as with book libraries we expect to find extensive holdings of journals and books, many of which have never been checked out but are ready for use at any time.

It is likely that usage monitoring will also reveal that most users of mathematical software could be classified as small or casual users, rather than groups who develop and maintain large modeling and simulation codes. Again, one should not be surprised by such data. Casual users are not likely to develop very good mathematical routines of their own in the limited amount of time they can devote to programming. Thus, they may well need libraries more than groups who develop large codes and have full-time programming staffs. Furthermore, users with small programs do not always have the fanatical interest in speed that developers of the largest codes usually have. Since small programs do not consume large amounts of computer resources, and therefore money, the usual fear of inefficiency in library routines does not apply here as forcefully.

In interpreting the data, one must remember that different computing styles will affect the usage counts revealed by monitoring techniques. Some operating systems encourage recompilation and loading every run, whereas others lead users to incorporate the library subprograms into their own files, instead of accessing the library every time. Since usage monitoring schemes typically only count the times a routine is loaded from a system library, in the latter case one does not get an accurate picture of the level of usage of the routines.

Finally, one must not expect much praise to be volunteered by the users. Scientists and engineers understandably have more interest in the phenomena they are modeling than in the computer model which they use.

D. Uses of the Data

While usage information is inaccurate in many ways and difficult to interpret, it can be used in several important ways:

- If users are still using obsolete routines, they can contact the individual users, make them aware of the shortcomings of these routines, and offer to help them select better ones.

- If input argument values are sampled, this information can be used to tune algorithms to local usage patterns.

- If a lot of cpu time is spent in a routine, then one can spend some time trying to make the routine more efficient for the local computing environment.

• If there are sufficient data, they can be used for cost/benefit analyses to justify more or less effort to be spent on mathematical software libraries.

• If bugs are found in routines, one can advise users of those specific routines that there is a bug.

Conclusion

This paper sets down some of the issues involved in supporting mathematical software libraries and describes procedures that have been used with moderate success. As is often the case in human relations, the attitudes and personalities of the people involved have as much influence on the effectiveness of the service as the procedures one defines.

References

[1] Smith, B. T., et al., "Matrix Eigensystem Routines - EISPACK Guide," Lecture Notes in Computer Science, Vol. 6, 2nd Edition, Springer-Verlag, 1976.

[2] Cody, W. J., "The FUNPACK Package of Special Function Subroutines," TOMS, Vol. 1, March 1975, pp. 13-25.

[3] Dongarra, J. J., Bunch, J. R., Moler, C. B., and Stewart, G. W., LINPACK Users' Guide, SIAM, Philadelphia, 1979, C. 112.

[4] More, J. J., Garbow, B. S., and Hillstrom, K. E., "User Guide for MINPACK-1," Argonne National Laboratory Report ANL-80-74, August 1980.

[5] Swarztrauber, P. Sweet, R., "Efficient FORTRAN Subprograms for the Solution of Elliptic Partial Differential Equations," NCAR Technical Note NCAR-TN/IA-109, July 1975.

[6] Bailey, C. B. and Jones, R. E., "Usage and Argument Monitoring of Mathematical Library Routines," ACM Transactions on Mathematical Software, Vol. 1, No. 3, September 1975, pp. 196-209.

[7] Chan, T. F., Coughran, W. M., Jr., Grosse, E. H., and Heath, M. T., "A Numerical Library and Its Support," ACM Transactions on Mathematical Software, Vol. 6, No. 2, June 1980, pp. 135-145.

Appendix A

Format of AMDLIB Subroutine Documentation

The following format is suggested for AMDLIB subroutine documentation. Deviations from this format should be discussed with the AMDLIB Librarian.

ARGONNE NATIONAL LABORATORY
APPLIED MATHEMATICS DIVISION

A statement of the purpose of the program
(no more than three sentences)

AMDLIB I.D. number   Member name

Author and/or AMD sponsor

Date

I. Purpose

II. Usage

- A. Entry Point
- B. Calling Sequence
- C. Organization of Routine
- D. Error Conditions and Returns
- E. Applicability and Restrictions
- F. Data Format

III. Discussion of Method and Algorithm

- A. Method or Example
- B. Range and Domain
- C. Error Analysis

IV. References

V. Program Statistics

A. Non-standard Routines Required

B. Common Storage

C. Space Requirements

VI. Program Materials Available

A. Writeup - number of pages

B. Source Listing - number of pages

C. Compiler Listing - number of pages

D. Source Deck - number of cards

E. Object Deck - number of cards

F. Test Decks

Test No. 1 - number of cards

Test No. 2 - number of cards

G. Test Listings

Test No. 1 - number of pages

Test No. 2 - number of pages

H. Demonstration Deck - number of cards

I. Demonstration Listing - number of pages

VII. Checkout

A. Accuracy Testing

B. Timing Testing

C. Error Return Testing

D. Demonstration Programs

VIII. Sample Input and Output

A. Sample Problem

B. Sample Input

C. Sample Output

IX. Reasons for Revision (if applicable)

X. Source Listing

NOTES ON DOCUMENTATION FORMAT

Identification of Routine

A. The AMDLIB I.D. number begins with a letter indicating the classification of the subroutine. The next three digits identify the sub routine within its classification. The next character is an S indicating compatibility with the IBM 360 system. The sixth character indicates the language:

A: Assembler language.
F: Fortran.
P: PL/I

The final digit indicates the revision number.

B. The member name is the name of the main entry point of the routine (e.g., the subroutine name).

C. If the author is not from ANL, the author's name and institution are stated, then the AMD sponsor's name follows.

D. The date is the time when the routine was added to the library. Following this are the dates when any modifications were made.

I. Purpose

One or two sentences describing the intended purpose of the program.

II. Usage

A. The names of the entry points in the subroutine.

B. The call must contain a sample calling sequence for each entry point (two or more may be illustrated by the same example). For Fortran-compatible routines, the corresponding Fortran calling sequence must also be given. Parameters that are supplied to the routine must be listed together with a full description of their purpose and format.

C. If there are other subroutines used internally by the main subroutine, these subroutine names and functions must be briefly described. The control and calling sequences of these routines must be stated clearly. Routines that must be supplied by users have to be fully described.

D. Error conditions and returns should list each situation that will result in an error return, together with an explanation of how and where the return is made.

E. Any restrictions such as machine hardware features, compiler options required, labeled common areas, etc., must be listed. If Assembler language is used, information about the register usage, such as which registers are used, which are saved and restored, and which are destroyed by the routine, must be listed.

F. This section should give an adequate description of the input/output devices, DCB information on datasets, etc.

III. Discussion of Method and Algorithm

A. State the method used or give an example to illustrate the method.

B. Range and domain must give the limitations of the routine, such as: the range of the arguments, number of equations that can be solved, limitation on results returned, etc.

C. Error analysis must contain an analytic discussion of the error to be expected. A reference to a publication which clarifies or describes the analysis would be helpful.

IV. References

Books, papers, or any other literature referenced in the documentation must be listed.

V. Program Statistics

A. Routines that are required from AMDLIB or supplied by the user should be listed and fully described.

B. Common storage. The name and size of each (labeled) common area must be listed. A brief description of each variable in the common areas must be given.

C. Space requirement. Give the number of bytes the routine occupies in hexadecimal and in decimal. Space required by the common areas must be included. The compiler used and a release number must also be listed.

VI. Program Materials Available

A list of the various card decks, computer output listings, tapes, writeup, and materials pertaining to the subroutine.

VII. Checkout

A. Test cases must describe specific examples to validate the results claimed for accuracy. Whenever possible, tables of the testing results should be presented.

B. Timing information of the test cases must be given. The model of the machine used in the tests and the compiler options used in the routine being tested must be stated.

C. Error test cases must test all possible error returns.

D. The demonstration program must be easy to understand and must be self-checking.

VIII. Sample Input and Output

A. The statement of the sample problem, the function of the sample problem, and the output format must be stated. Output must not be longer than 80 characters per line.

B. The sample input is the source listing of the demonstration program.

C. The sample output is the output of the demonstration program.

Appendix B

Mathematical Notation in Machine-Readable Form

The documentation of routines for AMDLIB must be prepared in machine-readable form. It is possible to represent mathematical notation in machine-readable form. However, there are cases where the representaton of a mathematical statement could be very messy. Therefore, detailed mathematical analysis should not be included but instead pointed to in references such as books, papers, or technical memos. The following are the machine-readable forms of commonly used mathematical symbols; if additional ones are needed, this list will be extended.

1. Logical Operations

Notation	Machine-Readable Form
$\geq$	.GE.
$>$	> or .GT.
$=$	= or .EQ.
$<$	< or .LT.
$\leq$	.LE.
¬=	.NE.
$\wedge$	.AND.
$\vee$	.OR.
¬	.NOT.

2. Mathematical Constants

Notation	Machine-Readable Form
π	PI ; PI=3.141592653...

3. Mathematical Operations

Notation	Machine-Readable Form
+	+
-	-
x	*
÷	/
power	**
$\sum_{i=1}^{n} a_i$	SUM A(I) (from I=1 to N) or SUM (I from 1 to N) (A(I))
$\prod_{i=1}^{n} a_i$	PRODUCT A(I) (from I=1 to N) or PRODUCT(I from 1 to N) (A(I))
$\oplus$	DIRECT SUM
n!	PRODUCT (I) (from I=1 to N) or Factorial (N)
$\leftrightarrow$	Iff or If and only if
$\rightarrow$	Implies
$\overline{\lim}$	Lim Sup
$\underline{\lim}$	Lim Inf
$f : A \rightarrow B$	f is a mapping from set A to set B

4. Greek Letters

λ	LAMBDA
α	ALPHA
β	BETA

5. Mathematical Functions

For mathematical functions, we prefer the function entry names (single-precision) used by the Fortran language:

Function Name	Standard Notation	Machine-Readable Form
Natural log,	$\ln(x)$	LOG(X)
Common log,	$\log(x)$	LOG10(X)
Exponential,	e^x	EXP(X)
Square root,	$\sqrt{x}$	SQRT((X)
Arcsin,	$\sin^{-1}(x)$	ASIN(X)
Arccos,	$\cos^{-1}(x)$	ACOS(X)
Arc tangent,	$\tan^{-1}(x)$	ATAN(X)
Sine,	$\sin(x)$	SIN(X)
Cosine,	$\cos(x)$	COS(X)
Tangent,	$\tan(x)$	TAN(X)
Cotangent,	$\cot(x)$	COTAN(X)
Hyperbolic sine,	$\sinh(x)$	SINH(X)
Hyperbolic cosine,	$\cosh(x)$	COSH(X)
Hyperbolic tangent,	$\tanh(x)$	TANH(X)
Absolute value	$\|x\|$	ABS(X) or |X|
Error function		ERF(X)
Gamma function	$\Gamma(x)$	GAMMA(X)
Log gamma	$\ln \Gamma(x)$	LGAMMA(X)
Modulo arithmetic,	$z \equiv x (\bmod y)$	Z=MOD(X,Y)

6. Matrix Notation

Let A be a matrix.

Inverse of A	A^{-1}	INVERSE(A)
Transpose of A	A^T	TRANSPOSE(A)
Diagonal matrix with array W as diagonal elements		DIAG(W), W=W(I,I)
Determinant of A		DET(A)
Trace of A		TRACE(A)

7. Mathematical Notation

$\int_a^b f(x)\,dx$	INTEGRAL(FROM A TO B) F(X) DX.
$\frac{dy}{dx}$	dY/dX or Y'
$\frac{d^2y}{dx^2}$	(d/dX)(dY/dX) or Y''
∞	INFINITY
$-\infty$	- INFINITY
$\in$	is an element of
$\subset$	is a subset of
$\exists$	there exists
$\frac{\delta f(x,y)}{\delta x}$	(d F(X,Y)/dX)
$\frac{\delta^2 f(x,y)}{\delta x \delta y}$	(d/dX)(dF/dY)

Appendix C

Sample Mathematical Routine Documentation

To illustrate the effort devoted to documentation, in this appendix we reproduce the write-up for one subroutine in the Argonne library. Note the advice on when to use the routine, how to determine whether the results can be trusted, and the extensive checkout of timing and accuracy.

ARGONNE NATIONAL LABORATORY

APPLIED MATHEMATICS DIVISION

SYSTEM/360 LIBRARY SUBROUTINE

ANL D452S ENDACE

EVALUATE NUMERICAL DERIVATIVES AND CORRESPONDING ERRORS

J. N. LYNESS

AUGUST, 1976

I. PURPOSE.

A subroutine for numerical differentiation at a point. It returns a set of approximations to the j-th order derivative (j=1,2,...14) of FUN(X) evaluated at X = XVAL and, for each derivative, an error estimate (which includes the effect of amplification of round- off errors).

II. USAGE.

A. Calling Sequence.

```
IMPLICIT REAL *8(A-H,O-Z)
EXTERNAL FUN
DIMENSION DER(14),EREST(14)
CALL ENDACE(XVAL,NDER,HBASE,DER,EREST,FUN)
```

1. Input Parameters:

XVAL Double precision. The abscissa at which the set of derivatives is required.

NDER Integer. The highest order derivative required.
If(NDER.GT.0) all derivatives up to MIN(NDER,14) are calculated.
If(NDER.LT.0 AND NDER EVEN) even order derivatives up to MIN(-NDER,14) are calculated.
If(NDER.LT.0 AND NDER ODD) odd order derivatives up to MIN(-NDER,13) are calculated.

HBASE Double precision. A step length.

FUN The name of a double precision function subprogram which is required by the routine as a subprogram and which represents the function being differentiated. The routine requires 21 function evaluations FUN(X), located at X = XVAL and at

X = XVAL + (2*J-1)*HBASE, J=-9,-8, ... +9,+10

The function value at X = XVAL is disregarded when only odd order derivatives are required.

2. Output Parameters:

DER(J) J=1,2,...14. Double precision. A vector of approximations to the J-th derivative of FUN(X) evaluated at X = XVAL.

EREST(J) J=1,2,...14. Double precision. A vector of estimates of the absolute accuracy of DER(J). These are negative when EREST(J).GT.ABS(DER(J)), or when for some other reason the routine is doubtful about the validity of the result.

B. Organization of Routine.

The only component of this routine is the subroutine ENDACE which controls the overall differentiation of a real valued function at a point.

FUN(X) is a user supplied double precision function subprogram which represents the function to be differentiated.

C. Error Conditions and Returns.

The sign of EREST is an error indicator. EREST(J) is set negative if the corresponding calculated derivative DER(J) is grossly inaccurate. This happens either if EREST(J) exceeds DER(J) in magnitude, or if two previous consecutive erests of the same parity are negative.

Note: EREST is an estimate of the overall error. It is essential for proper use that the user checks each value of DER subsequently used to see that it is accurate enough for his purposes. Failure to do this may result in the contamination of all subsequent results. It is to be expected that in nearly all cases DER(14) will be unusable. (14 represents a limit in which for the easiest function likely to be encountered this routine might just obtain an approximation of the correct sign.)

III. DISCUSSION OF METHOD AND ALGORITHM.

The method and theory used in ENDACE are documented in detail elsewhere (1), and a description of the extended t-table (TSUB,K,P,S), may also be found (2,3).

The problem to be solved is to find a set of approximations, DER(J), J=1,2, ... ,N , to the Jth derivative of a real valued function, FUN(X), at a real valued abscissa, XVAL.

A set of error estimates, EREST(J) J=1,2, ... ,N , is also found where EREST(J) is an estimate of the absolute accuracy of DER(J).

Note: In the case in which the user can code a complex function FUN(X), S = X+SQRT(-1.0)*Y, there is a more reliable algorithm available (but not in AMDLIB yet), namely CACM algorithm 413 (ENTCAF/ENTCRE). This uses complex function values to determine real derivatives, thus avoiding round-off error amplification problems. (See reference 4).

A. Differentiation Method.

The derivative approximations are based on 21 function values, FUN(X), located at:

X=XVAL, X=XVAL + (2*J - 1)*HBASE, J=-9,-8, ... ,+9,+10.

Internally, the routine calculates the odd order derivatives and the even order derivatives separately. For each derivative, the routine employs an extension of the Neville algorithm (3), to obtain a selection of approximations.

For example, for odd derivatives, based on 20 function values, the routine calculates a set of numbers:

T(K,P,S)   P=S,S+1, ... ,P*   K=0,1, ... ,K*-P

with   P* = 6   and   K* = 9

each of which is an approximation to the (2S+1)th derivative of FUN(XVAL)/(2S+1)! A specific approximation T(K,P,S), is of polynomial degree 2P+2 and is based on polynomial interpolation using function values FUN(XVAL+(2I-1)H) , FUN(XVAL-(2I-1)H) I=K,K+1, ... ,K+P. In the absence of round off error, the better approximations would be associated with the larger values of P and K. However, round off error in function values has an increasingly contaminating effect for successively larger values of P. This routine proceeds to make a judicious choice between all the approximations in the following way. These approximations may be classified into (P*-S+1) sets, S(P,S), (S.LE.P.LE.P*).

Denote:

A(2S+1) = the (2S+1)th derivative of FUN(XVAL)/(2S+1)!

A(2S+2) = the (2S+2)th derivative of FUN(XVAL)/(2S+2)!.

B(S) = A(2S+1) or A(2S+2), depending on the order of the derivative.

Each set S(P,S) contains the approximations to B(S) having the same P-value (or polynomial degree). Thus S(P,S) contains K*-P+1 approximations T(K,P,S) K=0,1, .. ,K*-P.

Now, for a specified value of S, a single value B*(S) and an error estimate E(S) must be chosen by the routine. In ENDACE a very simple algorithm is used to obtain B*(S) and E(S).

Denote the extreme elements in S(P,S) by:

MX(P,S) = MAX T(K,P,S), MN(P,S) = MIN T(K,P,S) over all K.

and define:

B(P,S)=(SUMMATION(T(K,P,S)-MX(P,S)-MN(P,S))) / (K*-P-1)
from K=0, to K=K*-P.

E(P,S) = MX(P,S)-MN(P,S)

When the set of different estimates B(P,S) and E(P,S) is available for each of the P*-S+1 sets S(P,S) (S.LE.P.LE.P*), the routine simply returns the result for which E(P,S) is smallest. Thus

E(S) = MIN E(P,S) = E(Q,S), over all P
B*(S) = B(Q,S).

Perhaps an example of the t-table calculated in the routine would prove helpful.

i) the table shows all the approximations to A(2) and some of the approximations to A(3).
ii) the set S(5,2) of approximations to A(2) is indicated by asterisks. Each element in this set is an approximation of the same polynomial degree.
iii) the routine ENDACE uses the greatest and least of these elements to define an error estimate E(5,2) and takes the average of the rest of these elements as an approximation B(5,2) to A(2).
iv) having carried out the same procedure for the sets S(2,2), S(3,2), S(4,2), S(5,2), and S(6,2), the routine accepts B(Q,2) as A(2) and E(Q,2) as E(2) where E(Q,2) is the smallest of E(2,2), E(3,2), E(4,2), E(5,2), and E(6,2).

Part of the extended Romberg table.

```
T(0,2,2)
          T(0,3,2)
T(1,2,2)            T(0,4,2)
          T(1,3,2)          *T(0,5,2)*
T(2,2,2)            T(1,4,2)            T(0,6,2)
          T(2,3,2)          *T(1,5,2)*
T(3,2,2)            T(2,4,2)            T(1,6,2)
          T(3,3,2)          *T(2,5,2)*
T(4,2,2)            T(3,4,2)            T(2,6,2)
          T(4,3,2)          *T(3,5,2)*
T(5,2,2)            T(4,4,2)            T(3,6,2)
          T(5,3,2)          *T(4,5,2)*
```

```
T(6,2,2)        T(5,4,2)
        T(6,3,2)
T(7,2,2)

        T(0,3,3)
                T(0,4,3)
        T(1,3,3)        T(0,5,3)
                T(1,4,3)        T(0,6,3)
        T(2,3,3)        T(1,5,3)
                T(2,4,3)        T(1,6,3)
        T(3,3,3)        T(2,5,3)
```

B. Error Analysis

Amplification of round off error is an integral part of the calculation, but its effect is measured reliably and automatically by the routine at the time of calculation. In general, the amplification of small errors in function values overwhelms errors due to the subsequent usage of machine arithmetic.

IV. REFERENCES.

1) Lyness, J. N., Software for Numerical Differentiation, Available from the author, Applied Mathematics Division, Argonne National Laboratory.

2) Lyness, J. N. and Moler, C. B., "Van der Monde Systems and Numerical Differentiation," Num. Math., Vol. 8, 458-464, (1966).

3) Lyness, J. N. and Moler, C. B.,"Generalised Romberg Methods for Integrals of Derivatives", Num. Math., Vol 14, 1-14, (1969).

4) Lyness, J. N. and Sande, G.,"Algorithm 413 ENTCAF and ENTCRE (Evaluation of Normalized Taylor Coefficients of an Analytic Function)", Comm. A.C.M., Vol 14, 669-675, (1971).

V. PROGRAM STATISTICS.

A. Non-Standard Routines Required.

User supplied function subprogram FUN(X). (See II.B).

B. Space Requirements.

Size of program = C94 (hex) = 3220 (decimal) bytes under the IBM OS/360, level 21.7,FORTRAN IB - H OPT=2.

VI. PROGRAM MATERIALS AVAILABLE.

A. Writeup - 12 pp.
B. Source Listing - 5 pp.
C. Compiler Listing - 20 pp.
D. Source Deck - 269 cards.
E. Object Deck - 48 cards.
F. Test Deck - 510 cards.
G. Test Deck Listing - 8 pp.
H. Test Deck Output - 6500 cards.
I. Demonstration Deck - 122 cards.
J. Demonstration Deck Output - 687 cards.

VII. CHECKOUT.

A. Accuracy Testing.

This routine was tested on a variety of problems and on a large number of initial step sizes, HBASE. The following is a summary of the testing procedure routine and some sample output.

The testing program takes 30 seconds of processing time on an IBM 370/195 and produces 6,500 lines of output.

This testing program makes 1701 separate calls to endace using various functions and step lengths. The whole process can be repeated for different values of XVAL. In each of the 1701 cases ENDACE returns derivatives, DER(J), and corresponding error estimates EREST(J), J=1,2,...,14. The testing program also calculates the values of these derivatives EXACT(J), using analytic formulas.

Statistical summaries appear at various stages and at the end. In reading the final statistical summary the tester has an idea of how well or poorly ENDACE performs.

1. Definitions:

a) Noninformative Result:
EREST .GT. ABS(DER), this is indicated by a negative erest.

b) Informative Result:
Right If EREST .GE. ACTUAL ERROR.
Wrong If EREST .LT. ACTUAL ERROR. This is indicated by an additional entry, "F", in the right margin, and a negative sign in the EST REL ERROR column. An Informative answer can be wrong by a factor "F" = ACTUAL ERROR/EREST.

2. EXAMPLE 1. The following local output results from the function:

F(X) = 1/((X-ALPHA)**2 + BETA**2)
ALPHA = 0.40
BETA = 0.20
XVAL = 0.0
HBASE = 1/512 .

J	EXACT	DER(J)	EREST(J)	ACTUAL ERROR	EST. REL. ERROR	ACT. REL. ERROR
1	.20000D02	.20000D02	.2132D-13	-.107D-13	.107D-14	-.533D-15
2	.11000D03	.11000D03	.6082D-10	.328D-11	.553D-12	.298D-13
3	.72000D03	.71999D03	.2380D-08	.285D-08	-.331D-11	.396D-11
4	.49200D04	.49200D04	.1995D-04	-.249D-05	.406D-08	-.506D-09
5	.26400D05	.26400D05	.6140D-03	-.320D-03	.233D-07	-.121D-07
6	-.10440D06	-.10440D06	.6331D-01	.460D 00	.606D-04	-.440D-05
7	-.84672D07	-.84672D07	.1467D 03	.396D 02	.173D-04	-.468D-05
8	-.24172D09	-.24165D09	.1526D 07	-.716D 05	.631D-02	.296D-03
9	-.56537D10	-.56518D10	.2188D 08	-.187D 07	.387:-02	.330D-03
10	-.11737D12	-.12371D12	-.3025D 12	.633D 10	.258D 01	-.539D-01
11	-.20549D13	-.21547D13	-.2853D 13	.998D 11	.139D 01	-.486D-01
12	-.21169D14	.19031D16	-.1024D 18	-.192D 16	.484D 04	.909D 02
13	.50202D15	.24107D17	-.7752D 18	-.236D 17	.154D 04	-.470D 02
14	.47377D17	-.70793D21	-.2576D 23	.708D 21	.544D 06	.149D 05

For the same function and XVAL = 0.0, including results for 9 values of HBASE, namely:

HBASE = 1/2**K, K=4,5,..,12 the following statistics resulted:

NUMBER WRONG BY A FACTOR GREATER THAN	1.0	1.5	2.0	5.0
LOCAL RESULTS	1	0	0	0

TOTAL NUMBER OF RESULTS	(NCOUNT)	126
NUMBER OF NONINFORMATIVE RESULTS	(NONINF)	62
NUMBER OF INFORMATIVE RESULTS	(NINF)	64
NUMBER RIGHT	(NRIGHT)	63
NUMBER WRONG	(NWRONG)	1
PERCENTAGE OF INFORMATIVE RESULTS WRONG		1.6

For most similar examples there were no wrong results. This example is specially chosen to illustrate the sort of wrong result which may occur.

3. EXAMPLE 2.

The following statistics are the total statistics for 21 different functions:

F(X) = DEXP(ALPHA1*X) ALPHA1 = 1, 4, 16

G(X) = 1/((X-ALPHA)**2 + BETA**2) ALPHA = 0.0(0.1)0.5

BETA = 0.3, 0.2, 0.1

with 9 different points XVAL, and 9 different step lengths HBASE, namely:

XVAL = 0.0(0.07)0.56

HBASE = 1/2**K K=4,5,...,12 .

Making 21*9*9 = 1701 calls to ENDACE in total.

FINAL OVERALL STATISTICS:

Cumulative breakdown by order of derivative

ORDER,J	1	2	3	4	5	6	7	8	9	10	11	12	13	14
TOTAL	1701	1701	1701	1701	1701	1701	1701	1701	1701	1701	1701	1701	1701	1701
NONINF	133	142	284	259	336	454	616	838	922	1141	1177	1359	1403	1519

INFORM	1568	1559	1417	1442	1365	1247	1085	863	779	560	524	342	298	182
RIGHT	1565	1556	1416	1440	1365	1247	1079	860	763	557	497	337	278	171
WRONG	3	3	1	2	0	0	6	3	16	3	27	5	20	11
%WRONG/INF	.2	0.2	0.1	0.1	0.0	0.0	0.6	0.3	2.1	0.5	5.2	1.5	6.7	60

It can be seen that the higher order derivatives are much more difficult to obtain, as evidenced by the non-informative statistic. In addition they are of low accuracy, this accuracy being given by EREST. Again, care should be taken by the user in applying the results returned by ENDACE.

B. Time Testing.

Numerical differentiation is a useful procedure in that it requires a minimum amount of processing time. The run of the testing program made 1701 calls to ENDACE returning 14 derivative and error estimate pairs per call, and in total, only 17 seconds of CPU time were used on an IBM 370/195. This was using IV-H OPT=2. Though execution times will vary with the function used, ENDACE is a very economical routine.

VIII. SAMPLE INPUT AND OUTPUT.

A. Sample Input.

```
C
C    ****    ****    DEMONSTRATION PROGRAM    ****    ****
C
C  THE DEMONSTRATION PROGRAM CALLS ENDACE AND SUPPLIES PARAMETERS XVAL,
C  NDER, AND HBASE.
C
C
C  THIS DEMONSTRATION PROGRAM CALLS ENDACE 4 TIMES TO OBTAIN
C  APPROXIMATIONS TO THE FIRST FOUR ODD ORDER DERIVATIVES AT X = 0.5D0
C  OF A FUNCTION,
C         FUN(X) = DEXP(2.D0*(X-.5D0)).
C  THE TRUE DERIVATIVES ARE 2, 8, 32, AND 128 RESPECTIVELY.
C
C  THE 4 RUNS DIFFER BY USING DIFFERENT STEP LENGTHS:
C        HBASE = .3, .03,  003, AND .0003 .
C
C
```

```
      IMPLICIT REAL*8(A-H,O-Z)
      EXTERNAL FUN
      DIMENSION DER(14),EREST(14)
      DIMENSION TRUDER(14),ACTERR(14)
C
      WRITE(6,107)
C
C  THE TRUE DERIVATIVES ARE STORED IN THE ARRAY TRUDER.
      DO 1 K=1,7,2
    1 TRUDER(K) = 2.D0**K
C
C  SET THE PARAMETERS REQUIRED BY ENDACE.
      XVAL = 0.5D0
      NDER = -7
C
C  SET THE PARAMETER HBASE SO IT HAS 4 DIFFERENT VALUES IN THE
C  FOLLOWING LOOP:  .3, .03, .003, AND .0003.
      KOUNT = 0
      DO 10 I=1,4
      KOUNT = KOUNT-1
      HBASE = 3.0D0*10.D0**KOUNT
C
      CALL ENDACE(XVAL,NDER,HBASE,DER,EREST,FUN)
C
C  PRINTOUT THE RESULTS OF ENDACE.
      WRITE(6,100) I
      WRITE(6,101) XVAL,NDER,HBASE
      WRITE(6,102)
      WRITE(6,103) (DER(J), J=1,7,2)
      WRITE(6,104) (EREST(J), J=1,7,2)
C
C  COMPUTE THE ACTUAL ERROR IN THE DERIVATIVE APPROXIMATION,
C  ACTERR(J) = DER(J) - TRUDER(J)
      DO 2 J=1,7,2
    2 ACTERR(J) = TRUDER(J)-DER(J)
C
C  PRINTOUT THE ACTUAL ERROR.
      WRITE(6,99)
      WRITE(6,106)
      WRITE(6,105) (ACTERR(J), J=1,7,2)
```

```
C
C  PRINTOUT GENERAL COMMENTS.
      GOTO(3,4,5,6),I
    3 WRITE(6,110)
      GOTO 10
    4 WRITE(6,111)
      GOTO 10
    5 WRITE(6,112)
      GOTO 10
    6 WRITE(6,113)
   10 CONTINUE
      WRITE(6,109)
      STOP
C
C
C  FORMAT BLOCK
   99 FORMAT('-','COMMENT:  EXACT DERIVATIVES ARE 2, 8, 32, AND 128.')
  100 FORMAT('-',10X, ///20X,            'RUN',I1,' OF ENDACE.')
  101 FORMAT('0','INPUT:',10X,'XVAL =',F6.3,10X,'NDER =',I3,10X,
     C'HBASE =',F10.8)
  102 FORMAT('-','OUTPUT:',5X,'J=1',14X,'J=3',14X,'J=5',14X,'J=7')
  103 FORMAT('0','DER(J)',3X,D12.6,4D16.6)
  104 FORMAT(' ','EREST(J)',1X,D12.6,4D16.6)
  105 FORMAT('0','ACTERR(J)',D13.6,4D16.6)
  106 FORMAT(' ','COMMENT:  ACTERR(J) = ACTUAL ERROR = EXACT ',
     C'DERIVATIVE-DER(J).')
  107 FORMAT('1','THIS DEMONSTRATION PROGRAM CALLS ENDACE 4 TIMES TO ',
     C'OBTAIN APPROXIMATIONS TO'/2X,'THE FIRST 4 ODD ORDER DERIVATIVES'
     C' AT X = .5 OF A FUNCTION ,'/10X,'FUN(X) = DEXP(2.D0*(X-.5D0)),'/
     C2X,'THE TRUE DERIVATIVES ARE 2, 8, 32, AND 128 RESPECTIVELY.'//
     C2X,'THE 4 RUNS DIFFER BY USING DIFFERENT STEP LENGTHS:'/
     C20X,'HBASE = .3, .03, .003, AND .0003.')
  109 FORMAT('-',1X///30X,'GENERAL COMMENTS'/1X,'(1)  THE QUALITY ',
     C'OF THE RESULTS DEPENDS ON THE STEP LENGTHS CHOSEN.  THE'/
     C7X,'OPTIMUM STEP LENGTH IS DIFFERENT FOR DIFFERENT PROBLEMS.  ',
     C'FOR THIS'/7X,'PROBLEM IT IS ABOUT 0.03.'// 1X,'(2)  IN ALL ',
     C'CASES EREST(J) GIVES A RELIABLE BOUND ON THE OF ',
     C'DER(J).'//1X,'(3)  IN CASES WHERE RESULTS ARE JUNK, EREST(J) ',
     C'IS SET NEGATIVE.  IT IS'/7X,'IMPERATIVE THAT THIS IS CHECKED ',
     C'IN THE USERS PROGRAM.  OTHERWISE'/7X,'HE RISKS USING JUNK AS ',
```

```
     C'IN THE RUN1 RESULTS ABOVE.'///20X,'END OF DEMONSTRATION ',
     C'PROGRAM OUTPUT')
 110  FORMAT('-','COMMENT:  NOTE THAT EREST(J) IS NEGATIVE FOR ALL J,',
     C' INDICATING THAT THE '/11X,'RESULTS ARE **UNUSUABLE JUNK**.  ',
     C'THE REASON IS THAT HBASE IS'/11X,'TOO LARGE FOR THIS PROBLEM.')
 111  FORMAT('-','COMMENT:  THESE RESULTS ARE UNUSUALLY GOOD.  ',
     C'SEVEN FIGURE ACCURACY IS'/11X,'CLAIMED, AND OVER EIGHT FIGURES '
     C'ARE ATTAINED IN THE SEVENTH'/11X,'DERIVATIVE.  PROBABLY ',
     C'NEAR OPTIMUM VALUE OF HBASE.')
 112  FORMAT('-','COMMENT:  THESE RESULTS ARE USUABLE BUT NOT GOOD.  ',
     C'NOTE THE POOR ACCURACY'/11X,'OF DER(7).')
 113  FORMAT('-','COMMENT:  ONLY THE FIRST THREE RESULTS ARE USUABLE.',
     C'  THE FIRST IS VERY ACCURATE.'/11X,'THE FOURTH IS JUNK.  THIS ',
     C'IS A TYPICAL SITUATION WHEN HBASE IS TOO'/11X,'SMALL.  NOTE ',
     C'THAT THE VALUES OF EREST(J) CORRECTLY DESCRIBE THIS'/11X,
     C'SITUATION, GIVING RELIABLE BOUNDS.')
C
C
C  END OF DEMONSTRATION PROGRAM    ***   ***   ***   ***   ***   ***
      END
      FUNCTION FUN(X)
      IMPLICIT REAL*8(A-H,O-Z)
      FUN = DEXP(2.D0*(X-.5D0))
      RETURN
C  END OF FUNCTION SUBPROGRAM     ***     ***     ***     ***     ***
      END
```

B. Sample Output.

This demonstration program calls ENDACE 4 times to obtain approximations to the first 4 odd order derivatives at X = .5 of a function,

FUN(X) = DEXP(2.D0*(X-.5D0)).

The true derivatives are 2, 8, 32, and 128 respectively.

The 4 runs differ by using different step lengths:

HBASE = .3, .03, .003, AND .0003.

RUN1 OF ENDACE.

```
INPUT:          XVAL = 0.500          NDER = -7          HBASE =0.30000


OUTPUT:     J=1              J=3              J=5              J=7

DER(J)    0.227842D 01     0.621607D 01     0.462556D 02     0.983877D 01
EREST(J) -.118080D 02    -0.445822D 02    -0.218271D 03    -0.113257D 04
```

COMMENT: EXACT DERIVATIVES ARE 2, 8, 32, AND 128.

COMMENT: ACTERR(J) = ACTUAL ERROR = EXACT DERIVATIVE-DER(J)

```
ACTERR(J)-0.278421D 00     0.178393D 01    -0.142556D 02     0.118161D 03
```

COMMENT: NOTE THAT EREST(J) IS NEGATIVE FOR ALL J, INDICATING THAT THE RESULTS ARE **UNUSABLE JUNK**. THE REASON IS THAT HBASE IS TOO LARGE FOR THIS PROBLEM.

RUN2 OF ENDACE.

```
INPUT:          XVAL = 0.500          NDER = -7          HBASE =0.03000


OUTPUT:     J=1              J=3              J=5              J=7

DER(J)    0.200000D 01     0.800000D 01     0.320000D 02     0.128000D 03
EREST(J) 0.253131D-13     0.141293D-10     0.124023D-07     0.126889D-04
```

COMMENT: EXACT DERIVATIVES ARE 2, 8, 32, AND 128.

COMMENT: ACTERR(J) = ACTUAL ERROR = EXACT DERIVATIVE-DER(J)

```
ACTERR(J) 0.888178D-15     0.722311D-12    -0.682572D-09     0.639856D-06
```

COMMENT: THESE RESULTS ARE UNUSUALLY GOOD. SEVEN FIGURE ACCURACY IS CLAIMED, AND OVER EIGHT FIGURES ARE ATTAINED IN THE SEVENTH DERIVATIVE. PROBABLY NEAR OPTIMUM VALUE OF HBASE.

RUN3 OF ENDACE.

INPUT: XVAL = 0.500 NDER = -7 HBASE =0.00300

OUTPUT:	J=1	J=3	J=5	J=7
DER(J)	0.200000D 01	0.800000D 01	0.320000D 02	0.128119D 03
EREST(J)	0.448530D-13	0.378746D-08	0.366643D-03	0.205478D 02

COMMENT: EXACT DERIVATIVES ARE 2, 8, 32, AND 128.
COMMENT: ACTERR(J) = ACTUAL ERROR = EXACT DERIVATIVE-DER(J)

ACTERR(J)	0.208722D-13	-0.253016D-09	0.607304D-05	-0.119423D 00

COMMENT: THESE RESULTS ARE USABLE BUT NOT GOOD. NOTE THE POOR ACCURACY OF DER(7).

RUN4 OF ENDACE.

INPUT: XVAL = 0.500 NDER = -7 HBASE =0.00030

OUTPUT:	J=1	J=3	J=5	J=7
DER(J)	0.200000D 01	0.800000D 01	0.319689D 02	-0.407494D 05
EREST(J)	0.125455D-12	0.674825D-06	0.436893D 01	-0.426812D 08

COMMENT: EXACT DERIVATIVES ARE 2, 8, 32, AND 128.

COMMENT: ACTERR(J) = ACTUAL ERROR = EXACT DERIVATIVE-DER(J)

ACTERR(J) 0.546230D-13 -0.412317D-07 0.310806D-01 0.408774D 05

COMMENT: ONLY THE FIRST THREE RESULTS ARE USABLE. THE FIRST IS VERY A
THE FOURTH IS JUNK. THIS IS A TYPICAL SITUATION WHEN HBASE I
SMALL. NOTE THAT THE VALUES OF EREST(J) CORRECTLY DESCRIBE T
SITUATION, GIVING RELIABLE BOUNDS.

GENERAL COMMENTS

(1) The quality of the results depends on the step lengths chosen. The optimum step length is different for different problems. For this problem it is about 0.03.

(2) In all cases EREST(J) gives a reliable bound on the accuracy of DER(J).

(3) In cases where results are junk, EREST(J) is set negative. It is imperative that this is checked in the users program. Otherwise he risks using junk as in the RUN1 results above.

Panel Session on the Challenges for Developers of Mathematical Software

On the last day of the Seminar, a panel session was held on the general topic of mathematical software development. The session was chaired by Ing. P. De Meo. The panel members were the Seminar speakers: Cody, Dekker, Ford, Gentleman, Lyness, and Messina. A transcript of the discussion follows.

Ing. P. DE MEO

Ladies and gentlemen, I guess that now is my turn to thank Prof. Murli for giving Informatica Campania the opportunity to sponsor the seminar and me the privilege to act as chairman of this panel. Since our speakers and part of the audience do not necessarily know what the Italsiel Group, which includes Informatica Campania, does, I will say just a few words to illustrate our activity. The Italsiel Group since 1969 has focused its activity in the area of design and the development of electronic information systems. The Group, which by now employs roughly 1500 people, represents by far the largest computer consulting company in Italy and one of the largest in Europe. The main customers of Italsiel include, at the central public administration level, the National Accounting Office, Department of Treasury, Department of Education, and Department of Finance. At the local public administration level, Regional Governments such as the one in Friuli Venezia Giulia. Besides this activity our work includes a number of systems for major industrial and commercial companies. Naturally, within this kind of framework we develop some projects in the area of the mathematical software and we have a Group which started to work a few years ago in econometrics. We are very interested in seminars of this kind and this is the reason why we have tried our best to support the "International Seminar on Problems and Methodologies in Mathematical Software Production." At this stage I guess we can start asking our lecturers one provocative question: what has been the greatest success and the greatest failure in this area. Shall we start with Professor Lyness?

Prof. LYNESS

I expected to be last but one to be asked this question. This question was invented by Prof. Cody but he has refused to tell us the answer, so when he is asked remember that it is not fair because as he set the question he should know the answer. My answer is a personal one, based on conversations with my psychologist. As you know, when one becomes an American, as I did, one has to acquire a psychologist. I have a psychologist who advises me. He tells me that among the many, many problems I have that he has diagnosed, one important one is that of communication. He says that in my relationships with other people in the community I do not communicate properly; I do not tell people precisely what I want, and I do not take enough care to find out precisely what it is that they want. When I look at the general computing scene, when I look at the large amounts of money, time, and effort being spent in running small scale and large scale computations, I feel sometimes that everyone should heed the advice of my psychologist. I find at Argonne, for example, people are prone to use out-of-date methods of computation. They could cut their machine time by factors of possibly 100 simply by seeking the advice of experts who are in the same building. I feel that people should ask more questions and should consult more often to find out what the current situation is. The main general problem as I see it is to persuade the scientist or engineer who wants to do his problem quickly, that the time taken to get advice at the beginning of the problem is well worth it and may save later on a surgical operation to take one piece of code out and put another piece of code in. So I would like to answer both questions with a prayer. Let us work hard to solve this communication problem which is, at root, a human communication problem.

Prof. GENTLEMAN

When I started to think about this question I was also caused to think about a paper on program libraries that I submitted to a Conference in 1967, which was rejected on the grounds that program libraries are not of interest to any one and therefore the quality of the paper was irrelevant because the topic was boring. I think that the greatest success we have had in mathematical software has been the recognition by the user community that libraries do matter and that rather than everyone writing their own programs using the first idea that came into mind, that it makes such more sense to have high quality libraries, like the NAG Library, like the IMSL Library, which are widely available and which implement reliable quality software that enables people not to have to think in detail about how their computations are being done. So I think that is easily the greatest success and I think that it has been an enormous change over the period of the last 15 years. We have been successful in communicating with the user public. Where we have not been successful and where I think the greatest failure lies, is in persuading our colleagues who build hardware, who design programming languages and who implement them, that some of our needs differ from the first thing that came into their mind, which seems to be the way they insist on building hardware, defining programming languages and writing compilers. Two days ago, Brian Ford quoted a statement by John Rice that numerical computing amounts to 50% of all the computing that's done. I frankly believe that is an enormous exaggeration. I think the truth of the matter is relatively close to what the manufacturers, the hardware engineers and the language designers believe, namely that numerical computing is an irrelevant small part of the total computing business. But that is exactly why they have to let us do things our way because it is not going to interfere with anything else they want to do. The design of a floating point system on a machine, for instance, or behaviour of the floating point arithmetic in a programming language, are entirely irrelevant to the main thrust of what the manufacturers or the software houses are going to earn from their products. There is not a good reason for not letting us have things that at least allow us to get on with our job.

Prof. MESSINA

I would like to address a combination of the two questions. My view of the greatest success has been stated by Prof. Gentleman better than I could: The existence of libraries that are actually used is in itself a success. It has been very disappointing to me though that we and many people in computing spend so much time putting up with the shortcomings, the weak points of our tools, which includes the hardware, the operating systems, the compilers and so on. We almost make a career of learning the idiosyncrasies of each, so that we can produce a useful product. I recognize that ours is a very user oriented discipline and for that I am very happy. We don't build items of mathematical software just to make them beautiful and look at them, but rather because we want them to be used. However, when so much intellectual energy and creativity is expended on merely conforming to the problems of some design, I think it is harmful. I also have been disappointed that the general attitude has been that these are things that cannot be changed. Again I agree with Prof. Gentleman. It is an experimental science and yet in this case the experimental subject is open to change, it is not like the nature of the universe or of the atom. And yet, more often than not, we take the attitude that because it is already built, it is in use, that it cannot be changed, and we do not even attempt to. The greatest challenge that I see in the future, therefore, is to continue to be useful, to produce things that actually work and therefore take into consideration the tools we have, but to try very much to improve them. As personal computing becomes more prevalent, I think there will be a multiplicity of systems, even more than we have now, and I do not think we can afford to spend so much of our intellectual energy looking at what people have done right or wrong and just adapting to it. There will be too many combinations possible and I think we will go mad before we succeed.

Prof. CODY

In a sense this is an unfair question for me since, as Prof. Lyness has said, I posed it in the first place. But in the States, you see, when a professor does not know the answer to a question, he asks his students to work on it and then he takes credit for their answer. I have my opinions about what has been our greatest success and our greatest failure. I had hoped somebody on the panel would say that our greatest success has been the emergence of a discipline. Fifteen years ago, people who worked on numerical software (we did not have that term then), were simply programmers and it was considered, at least in the U.S.A., a waste of talent for a person with an advanced degree to do this type of work. I believe, through the efforts of many people and the success of many projects in the USA, that a discipline has emerged and now this is a recognized, professional area of activity. We have also had many, I would prefer to call them disappointments rather than failures. Certainly there are areas where much work is yet to be done. Communication problems are important but the greatest disappointment to me is a technical disappointment, a technical problem that I had thought we were ready to solve and apparently we are not. I had hoped that the performance profile approach of Prof. Lyness would be much more economical than it is. I had hoped that that particular approach to looking at the quality of software would be the key to distinguishing between good and bad programs, would help us to choose a program appropriate for a particular problem. I still hope that that approach can be studied and that we can find economical ways of using it, but I am disappointed that it is as expensive as it is at the moment.

Prof. FORD

a) The Greatest Success: Morven Gentleman identified the emergence of numerical software libraries as the greatest success in our area of activity during the past 15 years. I believe that it was the EISPACK project at Argonne National Laboratory that provided the turning point for numerical software. Prior to that, although many organizations and some voluntary groups had attempted to develop general and subject libraries, their impact was severely limited since they were available to and used by only a small number of people. Typically programmers only used numerical software developed in their immediate environment. Above all they must have a copy of the source text on cards! EISPACK was a carefully interpreted FORTRAN version of the excellent linear algebra algorithms of Wilkinson and Reinsch. A selected subset of the language was used by Brian Smith and his colleagues for reasons of transportability. The codes were tested properly and well documented. Wayne Cowell and others went to substantial efforts to validate the routines (through test sites) on different families of machines, and then announced the general availability of EISPACK. Slowly, first in the United States of America, next in parts of Europe and now I think throughout the world, there were programmers who looked at it, liked it, used it and then subconsciously accepted it as a standard for numerical software. The fundamental point was reached when users recognized that there were programmers, outside their immediate community, who could provide them with software that they could confidently use. EISPACK was generally available and used in all places with confidence, because it was quality numerical software. It was because of EISPACK, and possibly one or two other major software activities that we are here today.

b) The Greatest Failure: My greatest disappointment has been our profound inability to persuade the computer manufacturers to produce, and to sell to us, the equipment that we need. During the last 10 years the arithmetic provided has got measurably worse. Ten years ago there were computers, for instance, that provided correctly rounded arithmetic. There are very few machine families that provide such a facility now. The proposed IEEE standard for hardware arithmetic features is a shaft of light in an otherwise dark world. Unfortunately I believe it may be a very long time before we can persuade the manufacturers that it is in their interest as well as ours to provide facilities and features in hardware and compilation systems that enable users to easily move their software between different machine families. Too many of the manufacturers are still intent on tying communities, be it a local community (e.g. Oxbridge University Computing Service) or a technical community (e.g. Nuclear Physics) to a specific machine range by use of idiosyncracies and limitations within that specific machine range.

c) Some Challenges: How can I find out what users actually want, for example in the NAG Library? I believe that all I can do at the moment is to provide what

I perceive that they want. Is there a process, or a mechanism, which programmers will use to tell providers of software what they, the users, need?! Equally important, how can I make contact with the large reservoir of computer programmers who still do not use generally available software? Some argue that they cannot understand the documentation, others do not trust not-invented-here software, others simply do not know of the existence of such material, or care. The more we publicize our numerical software, the more existing users hear repeatedly about the material. The uninitiated remain unknowing. How can we reach these people? Sometimes one is told that the NAG Library software and documentation (and IMSL and EISPACK software and documentation) are "too oriented to the scientific community. The information is not ordered and expressed in a way that social and biological scientists can understand." Is there a manner of representation that would enable the whole community to effectively use the software? Finally I hope in the next five years that users (the scientific and technical community) can specify, design and implement a numerical computing language that meets our requirements, and we can then persuade the manufacturers to implement it in full on all their systems.

Prof. DEKKER

The first question, what is the biggest success, I think if you look what has happened in the past, 30 or more years of the computer age, then I think the biggest success has been the development of the hardware, the technology. This has made a tremendous impact of the computers in the world and computers are entering many aspects of life. On the other hand, if you look at the software, the situation is much less satisfactory. One is still speaking about the software crisis; this means that the software does not satisfy the specifications, it is not completed in time and things like that; very often it is also not reliable enough. I am not speaking particularly of numerical software, but more in general of software for applications and so on. If cars would be as reliable as software very often is, then we would not dare to drive these cars. Still people are buying computers with software which in many senses is unsatisfactory. So, the challenge is to overcome the software crisis. As I said before, computers are entering more and more in the public life, not only in the scientific area and in the administrative data processing, but also in many other areas. So we still have to do a lot of work; as you have heard this week, in the mathematical software area there is a lot going on and there has been a lot provided and a lot improved in the last 10 or 20 years. This has still to go on; we need better languages to enhance the reliability, to facilitate proving correctness, and things like that as I talked about during the seminar, and to facilitate documentation. This is also what Dr. Ford was mentioning: the documentation has to become suitable for a broader public. The machines have to become more suitable both for the public and for the software. There is another failure, which also my colleagues mentioned, that the arithmetic has even deteriorated instead of improved. I think a good thing that is now happening is the proposed IEEE standard for floating-point arithmetic. This is a very good standard for machine arithmetic, and I hope that this proposal will have an influence in the future not only for microprocessors but also for many other computers, so that the arithmetic will improve and will indeed get to a sufficiently high standard. Also I think in the future, if really the computer is going to be used by nearly everybody than it should be as close as possible to the people. In the future, the machines should therefore have base ten, optimal rounding, and sign-magnitude representation. I think I leave it at this.

CHAIRMAN

Thank you very much. Now that we have got six very articulate answers to the same question, I wonder if there are questions from the audience to the panelists.

AUDIENCE

I first would like to make a comment in relation to the statement that 50% of computing is numerical computing. There will be much more interest in knowing the nature of numerical computing with respect to the effect in terms of money that such a computation has. I mean how much money the computation requires to carry out. That was just a remark. What I would like to ask is from which area most of the people who now work in numerical software come from; for example, are they mostly engineers, because I would like very much to know how many mathematicians are attracted by numerical software.

Prof. MESSINA

The majority of those I know have a mathematics background as opposed to computer science, engineering, or physics.

Prof. FORD

Professor Murli commented earlier in the week that the British school of numerical analysis has always had a tendency towards the study of practical (algorithmic) problems as opposed to a theoretical (functional analysis) approach. This practical outlook has found natural expression in the development of numerical software. Certainly the vast majority of contributors to the NAG Library are mathematicians (or theoretical physicists) by education. This is equally true of contributors from Europe and North America. Those of us involved in the development, testing and implementation of numerical software do suffer intellectual hang-ups from time to time. The preparation of numerical software is not a science, it is an engineering discipline. It is concerned with methodology, with an evolving technology. You can be very close to the correct formulation of a methodology and fail totally to converge on the ideal approach, for reasons quite beyond your control. There is no simple metric by which you may judge your work. Yet those of us with backgrounds in mathematics or physics have been trained to work with precise predictions and measurements. You always know at the end of your labor whether you have succeeded or not. In numerical software there is no easy metric. Hence those who move from mathematics and from hard science to engineering suffer a substantial psychological and philosophical shock. Regarding the first question, John Rice in his paper for the NSF Cosers project commented that 50% of all U.S. computing, in monetary terms, was numerical computing. It is an interesting observation. Perhaps one of my American colleagues would care to comment upon it. Recently a representative of IBM advised me that 28% of world computing by monetary value was scientific and technical, which he considered to be numerical computing. There was significant numerical computing in other areas (e.g. medical statistics, econometrics). Hence 28% was a lower bound, and this percentage had grown annually for each of the last three years.)

Prof. MESSINA

May I comment on something Prof. Ford just said? You were talking about the engineering nature of our discipline. In addition, however, there is a need for abstraction, or at least the development of methodologies which has been neglected somewhat. In many of the papers this week we discussed methodologies, but I have the impression that many mathematical software projects still do not take the time to analyze, once a project is completed, what have been the problems, what have been the successes and try to derive methodology from them. We must do that as well as the engineering considerations.

Prof. GENTLEMAN

I would like to express a cynical comment in answer to the question of where the people come from. It has been my experience that you can answer that question by looking at the software. There is something in a way, at least in North America, that physicists are educated, that mathematicians are educated, and that engineers are educated and that leads to qualitatively different styles of programming. Mathematicians have trained into them at an early stage thoughts of rigor and elegance and the quality mathematical software that I know of has almost universally come from mathematicians because they are the only people who are prepared to spend time and energy to achieve that quality and it's done really largely because of the motivations of elegance, that in a second year calculus course get them extra marks if they can prove the result in three lines instead of fifteen. Physicists, on the other hand, somewhere along their education learn that if you put your head down and beat against a concrete wall long enough there will be a hole in the concrete wall eventually. And their programs tend to look at this. It turns out that in the case of many large scientific codes that have been written by physicists, they are very nearly unmaintainable, so no one understands what they did, or why, but they do get answers people are happy with; often no one knows whether that means the answers are right. It has been my experience that if you want to have answers rapidly you hire a physicist to write the program. They are amazing; they will produce working code for an enormously large project in much shorter time that anyone else will, but as I remarked the code has a very distinct style to it and there again I think it has something to do with the education process whereby the typical undergraduate physicist learning his elementary statics course discovers that if with 13 pages of trigonometry he can show that some monkey climbing a rope in fact goes up instead of down when the weight at the other end of the rope is moving against some pulley, that was the only thing anyone cared about. The fact that the result could have obtained in three lines and the fact that several of the lines in the middle of his 13 pages are "fudged" to get the right answer does not bother anyone. Finally the engineers intrigue me in terms of the quality of code that they write, because it tends to be quite different from the style of code that physicists write. The physicists actually set out and produce answers. I have seen large numbers of engineering projects where in fact they gave up and they do not actually attempt to get the computer to do all of the things that they originally set out to do. They certainly do not prove that the methods that they use work, but they have been remarkably creative at inventing methods that do in fact work, or at least work on the problems that they are interested in, and which other people can come along later and analyze to understand why they work and explore to find a domain of where they are useful. I have in mind, for instance, the fact that the entire subject of sparse matrices arose out of working engineering programs. Not out of mathematicians interested in it or out of people who were trying to solve a single specific problem, it arose out of a general need for that kind of work in engineering. If I was trying to dictate how people should be trained, I think I would believe that there ought to be more in the way of mathematical engineering; students who get the exposure to the real world that exists (and that engineers are exposed to) and students who get the taste of elegance and rigor which mathematicians are exposed to.

CHAIRMAN

I suppose that answers your question, doesn't it?

Prof. MURLI

Which are the prospects of mathematical software and numerical libraries for microcomputers?

Prof. CODY

Perhaps we should repeat the question for the recorder since Prof. Murli does not have the microphone. Did I understand you correctly, you wanted to know what are the prospects for mathematical software on microprocessors? I believe they are quite good. You must understand that microprocessors are so readily available now that people who are not familiar with what has been done on larger machines are purchasing them and using them in their own laboratories with little communication with the outside world. These people are content that they have a machine and a programming language; they gather their data, open a book of mathematics and write programs for various processes. They are not aware that there are problems associated with certain processes; they are not aware that the problems exist, let alone that many of them have already been solved. The provision of programs for microprocessors is an attractive field for people who are somewhat less talented than the people who are providing programs for the large computers. We have seen in the popular literature in the United States, journals like Byte that you can purchase at the corner drugstore, programs which claim to solve particular types of problems on these small machines. We have seen small companies emerge which for a fee will provide a small library for these machines. Some of these libraries are copied from libraries that were thrown away 10 years ago for the larger machines because they were inadequate. We have seen mathematicians and physicists acquire a small machine and recognize the opportunity to become private business men. They write their own libraries and sell them. The quality of programs currently available for microprocessors is generally quite poor. They are often 15 to 20 years behind the technology. Not all programs are that bad, but the ones that are most widely used and the ones that I am afraid are likely to be used for the next 5 to 10 years are of that quality. There is a real challenge here.

Prof. FORD

Perhaps I could answer this question from a purely NAG viewpoint. At the present time (November 1980) we are defining a subset library specifically for 8-bit micros. We are selecting algorithms for this library apt for this environment. One version of the library will be in FORTRAN. NAG prepares libraries in FORTRAN, Algol 60 and Algol 68. However one of the most widely available languages on 8-bit micros appears to be Pascal. Some three years ago members of the Pascal language community suggested that NAG prepare a numerical library in Pascal. We pointed out that the absence of an internationally agreed standard for the language was a major disincentive to such an activity. Further the widely available Pascals had certain fundamental inadequacies that made the development of numerical software impossible (no library incorporation mechanism, no variable-dimensional arrays, no separate compilation, limitations on the use of procedures and functions as parameters). A proposed draft standard has now been prepared addressing the majority of these points and is being presented to the International Standards Organization for adoption. Even if this happens, Pascal dialects with the original deficiencies will continue to exist. We still believe that the translation of the full NAG Library into Pascal is neither a practical nor useful proposition. However to meet the substantial demand for Pascal-based software on 8-bit micros we are preparing a subset library in the extended Pascal. Microprocessors are being used in many interesting ways. In some universities programmers prepare their program on the mainframe, then feed it onto a micro and use it there for teaching or for research purposes. For the latter the machine may work overnight, and then be re-configured the next day to solve another problem. Used in this way much of our mainframe library software can be readily employed. For the micro market place, where the 8-bit micro is in stand-alone mode, we have to provide a different library service. The library will be on a disc or eventually on a chip (rather than a magnetic tape) prepared in read-only form. To stop plagiarism and theft we will provide only a pre-compiled library (no source text of routines or test software). Educationally and technically this is a retrograde step. Succinctly: NAG is designing a subset library of algorithms for use on 8-bit micros, in FORTRAN and Pascal. The service for stand-alone micros will be more limited than that for mainframes. However micros on networks will in general be able to load and execute any part of the full NAG Library, if they are specifically configured to do so. The 16 and 32 bit microcomputers with several hundred kilobytes of core, and 33 or 66 megabytes of filestore will, for numerical software, be directly comparable with present mini-computers. They present fascinating possibilities.

Prof. GENTLEMAN

I would like to have a chance to address Prof. Murli's question because when we started to go through the questions for the panelists I did not answer what is the most challenging problem to us but in fact my answer would have been that I think the microprocessor revolution does represent the greatest challenge to us. In answering I should perhaps define the context of what I think the microprocessor revolution means. If you look at the market for personal computers today, you will get a very poor prediction of what will be there in a year's time. The microprocessors which are prevalent today which are the heart of such personal computers as the Commodore Pet, or the Radio Shack TRS-80, or the Apple II are 8-bit machines that are very slow and horrible to program and for which many high level languages are used only interpretively or with very clumsy compilers. Within a very few months I believe that the 8-bit machine as the common thing you see, will have vanished and we will be instead talking about primarily microcomputers based on the 16-bit processors. This sounds like awfully low level discussion, but in fact there is a qualitative difference. Once you go to the 16-bit processors you can afford to compile good quality code, you can afford to have a high level language be the basis of all your software and you can afford to run large programs because the machines can run so much faster. In addition to the microprocessors having evolved, in the same time, we have had memory prices fall to the point where it makes no sense to have a personal computer that does not have a very large amount of memory. One firm that I am doing some contract work with is building a machine that will be shipped next year. We cannot afford on this personal computer to provide it in any configuration that has less than quarter of a million bytes of memory. It is too expensive to put small amounts of memory on machines and memory is cheap. The disk storage that comes with your personal computer, again it turns out that it makes no sense to have a small amount of disk storage. The price of the eight inch Winchester drivers is sufficiently low, especially when purchased in the volume that a manufacturer purchases them, that it is reasonable to expect all personal computers to have 20 to 40 megabytes of storage. What we are talking about then is a personal computer that you can expect every scientist, every engineer, every secretary to have on their desk. It is far more powerful than the large computers that many people had only a few years ago. Now I feel very enthusiastic about this because there are some interesting and perhaps unexpected side effects of all this. One of them is that the problem of program portability which we have had for the last 15 or 20 years is likely to go away with the advent of the microcomputers. That may surprise you but in fact the issue is that the hardware is so cheap and it changes so fast that the manufacturers of microcomputers systems cannot afford to develop the unique and monstrous software systems that we had to fight for the last 20 years. In the microcomputer business what is happening is that the manufacturers are standardizing on a few portable operating systems, on portable compilers, on things like standard libraries so they can get their products out. If you sit down with the new microprocessors, say the National Semiconductors 16000, which was announced a few weeks ago, and go to design a software system from its start, you are talking about a four- or five-year development project. If instead you pick up UNIX as an existing portable operating system, complete with compilers and useful commands, you can probably get the product out in six months. The person who makes the five year decision would be out of business long before he finishes his project. Now this is really exciting. It means that we have standard interfaces, we have very powerful machines and in that sense I think we are going to find that microprocessors are going to make our computing environment very much better. But there certainly are some problems associated with it and Brian Ford has touched one of the more painful ones. Writing software and distributing it has to be a commercial exercise. It costs money, we have to get money from somewhere, and it has to come from the users. We know how to build software projects and charge customers for it. When we are thinking of something like a large piece of software, that is sold to a computing center, that is sold to a firm that is going to use it, and we can control the distribution and make sure that people actually pay for a relatively easy way (because there is only one point of contact, there is only one computer on which it is going to run). We have absolutely no idea as a business proposition of how to deal with the problems of distributing hundreds of thousands of copies of software. We have no idea of how to market software where you have to charge perhaps $10 for a piece of software to an individual customer. The small shops that sell food, that sell records, that sell candy understand how to do those things. The computing industry does not. And I think that is going to be one of the most difficult problems to try to understand how to address and we all have the wrong training to do it.

Prof. MESSINA

I have not been reading Byte or Personal Computing very much myself. I got a subscription a few years ago and would only skim the issues. Finally, I stopped doing even that and cancelled the subscription. I feel guilty about that because it seems to me that some of the problems that Prof. Cody and Ford mentioned could be reduced in size if mathematical software people took more of an interest. I don't know for example whether anyone is writing articles in Byte that are at all comparable to Forsythe's paper on the Pitfalls of Computation. I think that paper by Forsythe along with a few other comparable ones made a big difference. They highlighted the discovery that it really did matter how one writes software to do numerical computations. If we are not contributing such articles to those journals that are read by the personal computing people then I think we have to take a lot of the blame if 20 year old methods in software are being used. Likewise we can be unhappy that programming languages have not evolved very much at all towards meeting the needs of numerical computation and yet until one year ago it has been many years that the Fortran standards committee in the U.S.A. had not had a single person on it who considered himself to be knowledgeable about numerical analysis or mathematical software. It is very difficult to change things if you don't take time to participate in the mechanisms that exist. Those mechanisms are unfamiliar to most of us. Also, we probably would not get a great deal of credit for publishing an article in Byte, as opposed to Mathematics of Computation. We don't get as much credit for being on the Fortran standards committee as for leading a Ph.D. dissertation I am sure. Yet if we want to make a difference in those fields I think we have to use the mechanisms that exist to change the situation.

CHAIRMAN

If there are no more comments on the side of the panelists, I wonder if there are any other specific questions from the audiénce, in terms of which is the major problem or one of the major problems that you think you are going to face in the near future.

Prof. E. RUSSO

I would like to return to the problem of software evaluation. To be precise, I feel that a strictly methodological approach to evaluation is of current interest because high quality routines for solving specific problems are being produced. Up to now the most significant and I think the only example is Prof. Lyness' "performance profiles". I would like to know what methodologies the panelists have used to evaluate routines up to now and what their plans are for the future, that is, whether work is underway at their institutions, NAG, Argonne, etc., to extend Prof. Lyness systematic methodology to other fields besides numerical integration.

Prof. LYNESS

At the moment very little has been done in making the performance profile approach more general. One of the difficulties is that people interested in testing the class two, the more complicated software, are usually the same people who wrote items of that software. In order to have the particular item of software they personally wrote accepted, they want to get comparisons quickly and they want the results of the comparisons to demonstrate that their own software is highly competitive or even better than its competition. A long time ago, I wrote a simple automatic quadrature routine called SQUANK. This was designed to demonstrate that the effect of round off error could be controlled automatically. It was fairly efficient but it was not a particularly good routine for doing quadrature unless round off error was expected to be significant. So people found, when they construct their new routines, that whatever happened it could always beat SQUANK in a competition. Thus it seems that every new routine published now compares the results of that new routine with SQUANK and the new routine is always better. SQUANK is very popular with routine constructors, but not with users. I feel strongly that in any examination or evaluation process it is important that people doing the evaluation should (1) know quite a lot about the subject and (2) not be personally involved with the results of the evaluation. And in many areas there are very few people around like that. Anybody who has taken the trouble to find out a lot about say minimization has written his own minimization routine. He hopes to get quick results so he can establish quickly that it is as good as or better than others. Testing routines other than one's own can be dull work. The person carrying out the testing may say to himself: "I am working hard to find out whether routine A is better than routine B. I'd much rather be finding new results for myself than finding whether routine A is better than routine B because I do not really care whether routine A is better than routine B or whether routine B is better than A. It may be helpful to others, but this result is not interesting to me." But it's work that should be done and it is rather tedious work. I think that the reason that this approach has not been developed very much, that is that we are all human beings. We don't want to do a lot of work which brings us personally little credit. To return to the question, as it affects me, I plan to start work on evaluation of optimization software during the summer of 1981. To force myself to do it, I have arranged to have a good student come to help me and I know that he will make me work on the project.

Prof. CODY

For once I disagree with Prof. Lyness. I think it is extremely important that the people developing tests have also developed software in the same area because software developers are the ones who know what the problems are and where the difficulties lie. (It is also extremely important that these individuals not have an ego, a sense of self importance.) They must recognize that whatever software they develop is limited in capability, that somebody else will write software that is better. It is a fact of life; and if they approach testing with the attitude that they are going to find that their program is not as good as they thought it was, then I think they can do a good job of writing test programs. I do agree with Prof. Lyness that it is a difficult task to find ways to do performance profiles. I think aside from Prof. Lyness' promise to work on this next summer, more correctly to have his student do the work, that the greatest hope we have is for people to think about the problem. There are many people at Argonne writing software for various purposes who are concerned about providing test material, about ways to check, to validate the software. We do talk to one another, believe it or not, and I have hope that one of these days the light will dawn, that in those conversations various ideas will come together and we will discover how to properly evaluate software. So we may not be working actively but we are thinking.

Prof. DEKKER

I would like to comment on this. In the first place, to evaluate a piece of software you have to make sure to have an insight in the structure of the algorithms which are used. As an example I mention what is going on in the area of optimization and solving systems of non-linear equations. There is a committee on algorithms of the Mathematical Programming Society which has the purpose to set standards for evaluating software for optimization and for solving systems of non-linear equations and they have made some publications about this. I mention Dembo and some others of the committee members who have worked out some proposals and this committee is still working on it. I would also like to mention the work of Bus who wrote a thesis this year which contains the results of extensive investigation of various algorithms (see reference below). What often happens in algorithms in this area, optimization and nonlinear equations, is that there are some constants chosen mainly by experience (for instance, a certain constant is 1/10 or 10**(-4) and it seems to work and things like that). What Bus did is, in each of these cases, try to find a mathematical theory in order to obtain some lower or upper bound for certain constants to make sure, or at least probable (in view of error estimates, etc.) that the algorithms work properly. Thus, he developed various new combinations or modifications of existing algorithms and compared the algorithms obtained with other existing algorithms. Moreover, he made a plan for comparing algorithms; in particular, he defined reliability criteria and robustness criteria, he did efficiency tests and so on. There came out two programs for general use which are suitable to put in a library. One program requires that the derivative, the Jacobian of the system is available (this is only on the non-linear systems problem), the other program does not use the Jacobian. There are some other programs which may be less suitable for general purpose libraries, but may be suitable for certain situations where you may prefer high efficiency at the cost of maybe somewhat smaller reliability or robustness.

Reference: J. C. P. Bus, Numerical solution of systems of nonlinear equations; Mathematical Centre Tracts 122, Amsterdam 1980.

Prof. GENTLEMAN

I was going to remark that in the last 15 years we have had a problem associated with testing that comes up with journals that have decided to publish algorithms or programs. In many cases the journals have expressed the desire to publish only programs that are of high quality and represent an improvement over previously existing programs in that area. In particular in the early 1970s, the late 1960s, Communications of the ACM in the Collected Algorithms tried very hard to do that. The effort caused enormous trouble and largely because of the question of testing that you asked us to respond to. It turned out that finding referees who are capable of doing a good quality comparison, of doing a good quality test, was extremely hard. The referees resented it because they get no credit for doing this work, their names did not even get published and those who were in the Universities and other places that require academic credit found that they were suffering; but perhaps the most serious problem was that the authors of algorithms that were being published resented the fact that it took maybe a year or two from the time that an algorithm was submitted until, even if it was better than all of other previous algorithms in that area, that fact could be established and then the algorithm published. Consequently, we had a change, the change happened now quite a number of years ago, but it was a noticeable change, that most journals that attempt to publish algorithms no longer make any effort to have these algorithms tested to see that they are even competitive with things that are already known.

Prof. LYNESS

I was worried to hear that Prof. Cody appeared to disagree with me. However, when he spoke I realized that it was another case of the communication problem, the human communication problem. I agree with every word he said. My comments were by way of example, representing what usually happens when I have tried to persuade other people to use the performance profile approach, people whom I respect, people who are clever and capable. What happens is roughly as I described it. The sort of person Prof. Cody described is hard to find. He has the character of a saint.

CHAIRMAN

No more questions? If not, I suppose Prof. Murli wants to close up the seminar.

ol. 107: International Colloquium on Formalization of Programming oncepts. Proceedings. Edited by J. Diaz and I. Ramos. VII, 478 ages. 1981.

ol. 108: Graph Theory and Algorithms. Edited by N. Saito and . Nishizeki. VI, 216 pages. 1981.

ol. 109: Digital Image Processing Systems. Edited by L. Bolc and enon Kulpa. V, 353 pages. 1981.

ol. 110: W. Dehning, H. Essig, S. Maass, The Adaptation of Virtual an-Computer Interfaces to User Requirements in Dialogs. X, 142 ages. 1981.

ol. 111: CONPAR 81. Edited by W. Händler. XI, 508 pages. 1981.

ol. 112: CAAP '81. Proceedings. Edited by G. Astesiano and C. Böhm. , 364 pages. 1981.

ol. 113: E.-E. Doberkat, Stochastic Automata: Stability, Nondeter- nism, and Prediction. IX, 135 pages. 1981.

ol. 114: B. Liskov, CLU, Reference Manual. VIII, 190 pages. 1981.

ol. 115: Automata, Languages and Programming. Edited by S. Even d O. Kariv. VIII, 552 pages. 1981.

ol. 116: M. A. Casanova, The Concurrency Control Problem for atabase Systems. VII, 175 pages. 1981.

ol. 117: Fundamentals of Computation Theory. Proceedings, 1981. dited by F. Gécseg. XI, 471 pages. 1981.

ol. 118: Mathematical Foundations of Computer Science 1981. oceedings, 1981. Edited by J. Gruska and M. Chytil. XI, 589 pages. 81.

l. 119: G. Hirst, Anaphora in Natural Language Understanding: Survey. XIII, 128 pages. 1981.

l. 120: L. B. Rall, Automatic Differentiation: Techniques and Appli- tions. VIII, 165 pages. 1981.

l. 121: Z. Zlatev, J. Wasniewski, and K. Schaumburg, Y12M So- on of Large and Sparse Systems of Linear Algebraic Equations. 128 pages. 1981.

l. 122: Algorithms in Modern Mathematics and Computer Science. oceedings, 1979. Edited by A. P. Ershov and D. E. Knuth. XI, 487 ges. 1981.

l. 123: Trends in Information Processing Systems. Proceedings, 31. Edited by A. J. W. Duijvestijn and P. C. Lockemann. XI, 349 ges. 1981.

l. 124: W. Polak, Compiler Specification and Verification. XIII, 9 pages. 1981.

. 125: Logic of Programs. Proceedings, 1979. Edited by E. geler. V, 245 pages. 1981.

l. 126: Microcomputer System Design. Proceedings, 1981. Edited M. J. Flynn, N. R. Harris, and D. P. McCarthy. VII, 397 pages. 1982.

l. 127: Y.Wallach, Alternating Sequential/Parallel Processing. 329 pages. 1982.

. 128: P. Branquart, G. Louis, P. Wodon, An Analytical Descrip- n of CHILL, the CCITT High Level Language. VI, 277 pages. 1982.

l. 129: B. T. Hailpern, Verifying Concurrent Processes Using mporal Logic. VIII, 208 pages. 1982.

l. 130: R. Goldblatt, Axiomatising the Logic of Computer Program- ng. XI, 304 pages. 1982.

l. 131: Logics of Programs. Proceedings, 1981. Edited by D. Kozen. 429 pages. 1982.

l. 132: Data Base Design Techniques I: Requirements and Logical uctures. Proceedings, 1978. Edited by S.B. Yao, S.B. Navathe, Weldon, and T.L. Kunii. V, 227 pages. 1982.

l. 133: Data Base Design Techniques II: Proceedings, 1979. ted by S.B. Yao and T.L. Kunii. V, 229–399 pages. 1982.

Vol. 134: Program Specification. Proceedings, 1981. Edited by J. Staunstrup. IV, 426 pages. 1982.

Vol. 135: R.L. Constable, S.D. Johnson, and C.D. Eichenlaub, An Introduction to the PL/CV2 Programming Logic. X, 292 pages. 1982.

Vol. 136: Ch. M. Hoffmann, Group-Theoretic Algorithms and Graph Isomorphism. VIII, 311 pages. 1982.

Vol. 137: International Symposium on Programming. Proceedings, 1982. Edited by M. Dezani-Ciancaglini and M. Montanari. VI, 406 pages. 1982.

Vol. 138: 6th Conference on Automated Deduction. Proceedings, 1982. Edited by D.W. Loveland. VII, 389 pages. 1982.

Vol. 139: J. Uhl, S. Drossopoulou, G. Persch, G. Goos, M. Dausmann, G. Winterstein, W. Kirchgässner, An Attribute Grammar for the Semantic Analysis of Ada. IX, 511 pages. 1982.

Vol. 140: Automata, Languages and programming. Edited by M. Niel- sen and E.M. Schmidt. VII, 614 pages. 1982.

Vol. 141: U. Kastens, B. Hutt, E. Zimmermann, GAG: A Practical Compiler Generator. IV, 156 pages. 1982.

Vol. 142: Problems and Methodologies in Mathematical Software Production. Proceedings, 1980. Edited by P.C. Messina and A. Murli. VII, 271 pages. 1982.